2nd International Symposium on Incoherent Light Sources

Enschede, Netherlands
9 – 12 April 1979

ISBN: 978-1-5108-4080-5

Printed from e-media with permission by:

Curran Associates, Inc.
57 Morehouse Lane
Red Hook, NY 12571

Some format issues inherent in the e-media version may also appear in this print version.

Printed by Curran Associates, Inc. (2017)

For permission requests, please contact Foundation for the Advancement of the Science & Technology of Light Sources (FAST-LS) at the address below.

Foundation for the Advancement of the Science & Technology of Light Sources
FAST-LS
Belmayne House
99 Clarkehouse Road
Sheffield, United Kingdom
S10 2LN

www.fast-ls.org

Additional copies of this publication are available from:

Curran Associates, Inc.
57 Morehouse Lane
Red Hook, NY 12571 USA
Phone: 845-758-0400
Fax: 845-758-2633
Email: curran@proceedings.com
Web: www.proceedings.com

2ND INTERNATIONAL SYMPOSIUM ON INCOHERENT LIGHT SOURCES

9–12 APRIL 1979
AT
TECHNISCHE HOGESCHOOL TWENTE
ENSCHEDE - NETHERLANDS

Information for Visitors

Summaries of Papers

2ND INTERNATIONAL SYMPOSIUM ON INCOHERENT LIGHT SOURCES

Technische Hogeschool Twente
Enschede – Nederland
9–12 April 1979

THIS IS THE SECOND SYMPOSIUM ORGANIZED TO BRING TOGETHER THOSE INVOLVED IN LAMP RESEARCH AND DEVELOPMENT THROUGHOUT THE WORLD, TO REPORT RECENT ADVANCES AND DISCUSS THE BASIC PHYSICAL AND CHEMICAL PROCESSES INVOLVED IN THE DESIGN AND OPERATION OF LIGHT SOURCES.

Secretariate:
Technische Hogeschool Twente
Department of Applied Physics – Building EF – Room 5270
P.O.Box 217 – 7500 AE ENSCHEDE the Netherlands
Telephone 053 – 89 34 25
Telex 44 200

CONTENTS

[1]) TWENTE is the name of the Eastern Part of the Netherlands Province 'Overijssel' = Trans Islia

1 SPONSORS

- National Research Council in the Netherlands (TNO)
- KEMA (Institute for the Quality Assurance of Electric Components and Equipment in the Netherlands)
- Ministry of Education and Science
- PHILIPS Research Laboratories Eindhoven
- Twente University Fund (Stichting Hogeschoolfonds Twente)
- Physics Foundation (Stichting Physica, founded in 1948)
- European Physical Society

2 ORGANIZING COMMITTEE

Dr.Ir. J.G.A.Hölscher	– Technische Hogeschool Eindhoven, NL.
Dr. M.Koedam	– PHILIPS N.V. Eindhoven, NL. —
Ir. A.H.C. van Meygaard	– Technische Hogeschool Twente, Enschede, NL.
Drs. E.W. ten Napel *	– Technische Hogeschool Twente, Enschede, NL.
Prof. J.F. Raffle, PhD.	– University of Technology, Loughborough, UK.
Prof.Dr. G. Szigeti **	– Hungarian Academy of Sciences, Hungary
Dr. J.F. Waymouth	– GTE. Sylvania Inc, Danvers Mass U.S.A.
Dr. D.O. Wharmby	– THORN Lighting, Leicester, UK.
Prof.Ir. H.G.de Winter	– Technische Hogeschool Twente, Enschede NL.

* *Acting Secretary during the months February, March and April 1979 THT - Building EF - 6th Floor - Room 6252. Telephone 053 - 893456.*

** *In January 1979 the committee received the message that Professor Szigeti died on the 26th of November 1978.*

3. GENERAL INFORMATION

3.1 HOW TO REACH TWENTE

By Air:

From Amsterdam Airport (Schiphol) a through-connection by plane exists except on Sundays.

From Twente Airport, the University, the ITC-guesthouse and hotels can be reached by taxi.

By Air/Train:

From Amsterdam Airport (Schiphol) a bus leaves for Amsterdam Centraal Station [note: *Central Station*] every 15 minutes, arriving at Amsterdam Centraal Station (CS) about 30 minutes later.

From Amsterdam–CS there is a fast train service every hour to Twente (Hengelo and Enschede). Look for destination ENSCHEDE. Visitors with destination Twente University have to travel to HENGELO. At HENGELO–station a bus leaves every 15 minutes to ENSCHEDE. Bus stop: TH–Twente.

Visitors with destination ITC–ENSCHEDE have to travel to ENSCHEDE. The ITC-Guesthouse is in the centre of ENSCHEDE (10 minutes walk or 5 minutes by taxi).

Visitors with destination DE LUTTE have to travel to OLDENZAAL*. The hotels can be reached by taxi (10 minutes).

By Boat/Train:

From 'Hook of Holland' boat-trains called Holland–Scandinavia/Hook–Warszawa/Nord-West/NoordZee Express leave at about half an hour after boat-arrival. These trains call at HENGELO. For further information see Air/Train.

By Car:

Coming from Utrecht/Deventer one has to drive first the way to HENGELO(O) (*not* to ENSCHEDE). In Hengelo choose the direction ENSCHEDE. The University is off the main road HENGELO–ENSCHEDE, and is well signposted on this road: "TH–TWENTE". Car-parking on the University ground is not limited.

3.2

TAXI	
ENSCHEDE	Tel. 300 000
HENGELO	Tel. 1 33 50
OLDENZAAL	Tel. 175 00

3.3 RESIDENTIAL ACCOMMODATION

Visitors will be housed in:
- one of the single study bedrooms on the campus
- a room of the ITC-Guesthouse, Enschede
- a hotel room in the surroundings.

* change at Hengelo.

Room Service

- campus : bed linen is provided; towels and soap are not provided and beds have to be made by the visitors.
- ITC-Guesthouse : *standard* hotel service.

Roomkeys

- campus : roomkeys are provided by the "Bureau Campushuisvesting", Vrijhof (map no. 2, building no. 22) room 128 during office-hours.
  In the week-end, please contact the staff at the main gate. (map no. 2, building no. 1).
- ITC-Guesthouse : roomkeys are provided at the ITC-desk in the Guesthouse. Open: 24 hours a day.

3.4 CANCELLATIONS

Participants are reminded that no payment can be refunded to those cancelling in whole or in part if such cancellations are received after March 25th.

3.5 WHAT TO DO ON ARRIVAL

If you arrive on or after 9th April, go straight to the Registration and Information desk in the hall next to the Conference Hall (map no. 2, building no. 2). This desk will be open from 08.30 hrs. on Monday, 9th April and troughout the rest of the conference.

If you arrive before Monday, 9th April, go straight to the ITC-Guesthouse or your hotel where further information will be available. If you are staying on the campus, collect your roomkeys from the staff at the main gate (map no. 1, building no. 1).

During specified hours on Saturday and Sunday someone of the organizing committee ('the attendant') will be present in the hall of the ITC-guesthouse (see 3.7).

Obtain your roomkey so that you can leave your luggage in your room, freshen up etc. If there are any complaints about your room please contact immediately the attendant. Registration can be made and if applicable, payment of the symposium-fee has to be made. Tickets for meals during the symposium are sold by the attendant (for prices see 3.11).

When you arrive before or during the weekend, the attendant can give you information on

- shopping
- theatres and movies
- divine services

PLEASE NOTE

In case of serious difficulties in the weekend or during the evenings please phone
Mr. E.W. ten Napel 05400 - 22139 or
Mr. A.H.C. van Meygaard 05410 - 15017.

3.6 HOW TO RECOGNIZE SYMPOSIUM-FACILITIES

Throughout the symposium the acronym ILS (Incohorent Light Sources) and/or the vignette as depicted on the cover of this symposiumbook will be used on signs referring to the conference hall, rooms, restaurants, bars, buses etc.

3.7 REGISTRATION TIME

	place	time
Saturday 7 April 1979	ITC-Guesthouse	17.00 - 18.00
Sunday 8 April 1979	ITC-Guesthouse	10.00 - 12.00
		19.00 - 22.00
Monday 9 April 1979	THT-Conference Hall	starting at 08.30

and further every day during the symposium.

3.8 REGISTRATION AND INFORMATION DESK

The desk will be open *all day* during the conference.

Telephone ENSCHEDE (053) – 893625

From	Germany (BRD)	00 3153893625
	Germany (DDR)	06 3153893625
	France	19 * 3153893625
	UK	010 3153893625
	Hungary	00 * 3153893625
	Japan	001 3153893625
	U.S.A.	011 3153893625

* wait for second signal

3.9 MEDICAL CENTRE / FIRST AID

- *Campus*: (On map no. 2, building no. 19): Medical Centre. Telephone: 2461. During weekends information will be given on tel. 2060.
- *ITC*: Hospital **STADSMATEN** (near ITC-guesthouse).

3.10 IN CASE OF FIRE

CAMPUS: immediately dial 2222
ITC: immediately dial 9 or 111

3.11 INFORMATION ABOUT MEALS

Tickets for meals can be obtained on arrival on the Registration desk in the ITC-Guesthouse (7 and 8 April) or at the Registration and Information desk (9, 10, 11 and 12 April).

Two types of tickets are available

1 A set of tickets for all lunches and dinners at Dfl. 100,–
2 A set of tickets for four lunches only at Dfl. 70,–.
THE SYMPOSIUM DINNER IS FREE

BREAKFAST ITC: Coffee-shop in the ITC-guesthouse between 7.30 and 9.00.
Campus: Breakfast will be served in the **Bastille Restaurant between 8.30 and 9.00.**

LUNCHES **are served in the** Bastille Restaurant (building no. 21) at
- 12.15 on 9th April
- 12.45 on 10th April
- 12.45 on 11th April
- 13.15 on 12th April

DINNER will be served in the **Bastille Restaurant** as follows
Monday: After the reception at the town-hall, dinner will be served at 19.00.
After dinner visitors can meet in the **Torenkamer-bar (Bastille)**; drinks are not free.
Tuesday: After the afternoon-session participants can meet in the **Torenkamer-bar.**
Dinner will be served at 18.30.

The symposium-dinner on wednesday 11th April will be served in the ITC-Guesthouse **(Schermerhorn-Hall)** Boulevard, Enschede.
The symposium-dinner is due at 20.00.

3.12 MAIL

All incoming mail should be clearly marked with the name of the symposium and will be given to the conference organizer for distribution.

3.13 OUTGOING TELEPHONE CALLS

CAMPUS: (white map) Telephone boxes - using 25-cents-coins **ONLY** - for outgoing calls are situated in
- the conference-hall (building no.2, ground floor, eastern entrance)
- the Vrijhof (building no. 22, ground floor, southern entrance).
- the BASTILLE (building no. 21).

ITC: (green map)
- Information-desk of the Guesthouse.

For long distance telephone calls, please *contact the*
1 **Porter-housing at the main-gate of the campus**
2 **Main post office, Enschede, Boulevard 1945 (five minutes walk from the ITC-guesthouse).**

To:

Germany (BRD)	09*49	Hungary	09*36
Germany (DDR)	09*37	Japan	09*81
France	09*33	U.S.A.	09*1
UK	09*44	* wait for second signal.	

3.14 MESSAGES / INCOMING TELEPHONE CALLS

During the day telephone messages and telegrams will be posted on the conference notice board in the lecture theatre concourse or given to the conference organizer for distribution.
The telephone number of the *information desk* will be

053 - 893625
from abroad: 3153893625

The *TELEX*-number of the University is:

44200

For sending out a *TELEX*-message, please contact the conference organizer.

Incoming telephone calls
Emergency only

Night	**ITC**	**053 - 312322**
	Campus	**053 - 892444**

3.15 CAMPUS SHOPPING CENTRE (map no. 2, building no. 44)

In the campus-shopping centre you will find:

a **Post Office (PTT)** Not for telephone calls (see 3.13)
The (small) post office is open during the following hours:

mo/tu/we/th/fr	9.00 - 13.30 h.
	14.30 - 17.30 h.
and mo/fr	17.30 - 19.00 h.

Postage Rates (January 1979)
Postcards: 40 cents.
Inland/France/BRD letters*: 55 cents
other countries letters: 75 cents
Air-letters (available at the post-office): 75 cents
Illustrated postcards are available at the post-office.
* max. 20 gr.
Letters by AIR: USA max. 10 gr.: Dfl. 1,–.

b **BANK (Algemene Bank Nederland N.V.)**
The University Branch of the 'Algemene Bank Nederland NV' is open at normal banking hours (9.00 - 16.00 h.).

c **Grocery shop ('de Spar') Supermarket**
The grocery shop/ supermarket 'de Spar' is open daily from 8.00 am. - 20.00 pm. (on Saturdays till 18.00 pm.).
Spirits and wines are available in the small shop 'Le Sapin' and this shop shows the same opening hours as 'de Spar'.

d **Self service laundry**
Between the post-office and 'de Spar' you will find a small self service laundry.

10

4. PROGRAMME

MONDAY 9 APRIL 1979

8.30 - 11.00 *REGISTRATION AND COFFEE*

11.00 *OPENING OF THE SYMPOSIUM*
– Prof.dr. I.W. van Spiegel; Rector Magnificus of the THT.
Chairman: – Prof. J.F. Raffle, PhD, on behalf of the organizing committee.

11.15 *INVITED PAPER*
LIGHT SOURCE TECHNOLOGY 1979 - A REVIEW, 22
Dr. J.F. Waymouth, GTE Sylvania Inc. Danvers, Mass. 01923, U.S.A.

12.15 *LUNCH IN THE BASTILLE-RESTAURANT (Campus - Centre)*

SESSION 1 HIGH PRESSURE DISCHARGES 1
Chairman: Dr. M.A. Cayless, Thorn Lighting Ltd., Leicester, U.K.

13.30 *INVITED PAPER* 24
THE DESIGN OF HIGH PRESSURE DISCHARGE LIGHT SOURCES.
Dr. R.J. Zollweg, Westinghouse R. & D. Center, Pittsburgh, Pennsylvania, U.S.A.

CONTRIBUTED PAPERS ON HIGH PRESSURE SODIUM (Na)-DISCHARGES

14.15 1 SOME IMPORTANT FACTORS IN DESIGN OF A COLOR IMPROVED HIGH PRESSURE SODIUM LAMP, 26
Ranbir S. Bhalla, Westinghouse Electric Corp., Bloomfield, N.J. 07003, U.S.A.

14.30 2 TOTAL EMITTANCE MEASUREMENT OF ARC TUBES IN HIGH PRESSURE SODIUM LAMPS, 28
E.F. Wyner, GTE Sylvania, Danvers, MA 01923, U.S.A.

14.45 3 DETERMINATION OF THE SODIUM AND MERCURY VAPOUR PRESSURE IN HIGH PRESSURE SODIUM LAMPS, 30
J.J. de Groot and J.A.J.M. van Vliet, Philips Lighting Division, Eindhoven, The Netherlands.

15.00 *TEA/COFFEE*
CONTRIBUTED PAPERS ON METAL HALIDE DISCHARGES

15.20 4 A NEW APPROACH TO THE DESIGN OF SODIUM/SCANDIUM HALIDE DISCHARGE LAMPS FOR GENERAL LIGHTING SERVICE, 32
S.A.R. Rigden, GEC Ltd., Hirst Research Centre, Wembley, Middx, U.K.

15.35 5 LOW POWER, HIGH PRESSURE, METAL HALIDE DISCHARGE LAMP, 34
K.E. Brown and A.G. Chalmers, Thorn Lighting Ltd., Research and Engineering Laboratories, Leicester LE4 7PD, U.K.

e **Book shop**

The book shop is housed in the Vrijhof (map no. 2, building no. 22, south-eastern part of the groundfloor and sells all kinds of books (also in English, German and French), stationery, magazines and newspapers.

The book shop is open during weekdays from 9 am. - 13.30 and from 14.00 - 17.00 pm.

f **University library**

The central university library is on the first floor of the building 'Vrijhof' (map no. 2, building no. 22, north-western part). The library is open for anyone during weekdays from 8.15 am. - 22.00 pm. In the library is also a newspaper-corner, where you can find newspapers as there are:

- Le Monde
- Le Figaro
- Daily Mirror
- Herald'Tribune
- The Observer
- The Guardian
- The Jeruzalem Post
- Die Zeit
- Neue Zürcher Zeitung
- Frankfürter Allgemeine

g **Copying Facilities for reproduction of papers(A4)**

In the central hall of the library a dry-copying-equipment* is available.
A 25-cents-coin is needed for 1 copy each.

* Rank Xerox 3100

PLEASE DO NOT FORGET TO RETURN YOUR KEY BEFORE LEAVING TWENTE

15.50 6 METAL HALIDE LAMPS WITH MOLECULAR RADIATION SPECTRUM, 35
G.S. Sarytchev, G.A. Naraikina, L.B. Prikupets, G.N. Gavrilkina, Moscow, USSR.

16.05 7 CHARACTERIZATION OF DIFFUSE AND CONSTRICTED HIGH PRESSURE ARCS BY MAGNETIC BALANCING, 36
P.C. Drop and W.J. van den Hoek, Philips Lighting Division, Eindhoven, The Netherlands.

16.20 8 A METHOD FOR DETERMINING THE TOTAL ACCUMULATED BURNING TIME OF LAMPS, 38
Vincent D. Meyer, Paul L. Fortucci and Elliot F. Wyner, GTE Sylvania, Danvers, MA 01923, U.S.A.

16.35 9 HIGH PRESSURE ELECTRODELESS MICROWAVE DISCHARGES AS LIGHT SOURCES, 40
Paul O. Haugsjaa, Robert J. Regan, William H. McNeil and Alfred E. Feuersanger, GTE Laboratories, Inc., Waltham, Mass. 02154, U.S.A.

16.50 *CLOSURE OF THE FIRST-DAY-SESSIONS*

17.00 *BUS DEPARTURE FROM THE LECTURE HALL FOR ENSCHEDE TOWN HALL*

17.30 *RECEPTION FOR SYMPOSIUM VISITORS BY THE BURGOMASTER OF ENSCHEDE, Mr. H. Wierenga*
To be held at the Town Hall, Langestraat 24, Enschede.

18.45 *BUS DEPARTURE FROM THE TOWN HALL FOR THE BASTILLE (Campus of the University)*

19.00 *DINNER AT THE BASTILLE-RESTAURANT*

20.00 *BAR AVAILABLE AT THE BASTILLE*

TUESDAY 10 APRIL 1979

SESSION 2 HIGH PRESSURE DISCHARGES 2

Chairman: Dr. J.F. Waymouth, GTE Sylvania Inc., Danvers, Mass. 01923, U.S.A.

9.00 *INVITED PAPER*
SYNCHROTRON RADIATION, 42
Prof.Dr. H.L. Hagedoorn, Technische Hogeschool Eindhoven, Dept. of Applied Physics, Eindhoven, The Netherlands.

CONTRIBUTED PAPERS ON MODEL STUDIES

9.45 10 CALCULATION OF RADIATION ENERGY TRANSPORT IN OPTICALLY THICK HIGH PRESSURE DISCHARGES, 44
B.F. Jones*, D.A.J. Mottram* and G. Eardley**
* Thorn Lighting Ltd., Research and Engineering Laboratories, Leicester LE4 7PD, U.K.
** School of Mathematics, Computing and Statistics, The Polytechnic, Leicester.

10.00 11 CALCULATIONS OF THERMAL CONDUCTIVITY AND OF THE EFFECTS OF RADIAL DIFFUSION OF ENERGY TRANSPORT IN HIGH PRESSURE DISCHARGES, 46
D.A.J. Mottram and B.F. Jones, Thorn Lighting Ltd., Research and Engineering Laboratories, Leicester LE4 7PD, U.K.

10.15 12 FREE CONVECTION AND QUARTZ WALL TEMPERATURES OF HORIZONTAL HIGH PRESSURE MERCURY ARCS, 48
D.K. McLain and R.J. Zollweg, Westinghouse R. & D. Center, Pittsburgh, PA, U.S.A.

10.30 13 EFFECT OF THE FREE CONVECTION ON THE TEMPERATURE DISTRIBUTION IN A WALL-STABILIZED HORIZONTAL ARC, 50
T. Vicsek, Institute for Technical Physics of the Hungarian Academy of Sciences, H-1325 Budapest, Hungary.

10.45 *COFFEE*

CONTRIBUTED PAPERS ON METAL HALIDE DISCHARGES

11.15 14 DIAGNOSTIC STUDY OF A DyI3/Hg-ARC PLASMA 51
W.J. van den Hoek and J.A. Visser, Philips Lighting Division, Eindhoven, The Netherlands.

11.30 15 STUDY OF THE Na-D AND THE Tl 535 nm-LINES EMITTED FROM A HIGH PRESSURE MERCURY DISCHARGE WITH SODIUM AND THALLIUM IODIDE AS ADDITIVES, 52
H.P. Stormberg, Philips GmbH Forschungslaboratorium - Aachen, Aachen, B.R.D.

15.45 *TEA/COFFEE*

CONTRIBUTED PAPERS ON LIGHT MEASUREMENT AND APPLICATION

16.15 23 THE PHOTOMETRY OF INCANDESCENT LIGHT-SOURCES; THE RECENT DEVELOPMENTS AND THE CHANGES TO THE PHOTOMETRY SI-UNITS IN THE NEAR FUTURE, 69

A.H.C. van Meijgaard, Technische Hogeschool Twente, Dept. of Applied Physics, Enschede, The Netherlands.

16.30 24 APPLICATION PARAMETERS OF A LIGHT SOURCE; LUMINOUS EFFICACY VS. COLOUR RENDERING, 71

J.J. Opstelten, Philips, Eindhoven, The Netherlands.

16.45 25 PRIME-COLOUR LIGHT SOURCES, 73

W.A. Thornton, Westinghouse Electric Corporation, Bloomfield, New Yersey, U.S.A.

17.00 26 MAPPING OF GaP LEDs PROPERTIES ON POLYCRYSTALLINE SSD SUBSTRATES, 75

D. Nohavica, Institute of Radioengineering and Electronics, Czechoslovak Academy of Sciences, Lumumbova 1, 18251 Praha 8, Czechoslovakia.

17.15 27 COLOR ELECTROLUMINESCENCE FROM ZnS THIN FILMS MULTI-DOPED WITH RARE-EARTH IONS, 76

S. Kurita, S. Takahashi, T. Sakaguchi, Faculty of Engineering, Keio University, Yokohama, Japan.

17.30 *CLOSURE OF THE SECOND-DAY-SESSIONS*
BAR AVAILABLE AT THE BASTILLE

18.30 *DINNER AT THE BASTILLE-RESTAURANT*

20.00 *BOWLING ALLEY AND BAR AVAILABLE AT THE BASTILLE*

WEDNESDAY 11 APRIL 1979
SESSION 4 INCANDESCENT LIGHT SOURCES

Chairman: Prof. E. Kauer, Philips Forschungslaboratorium Aachen, Aachen, B.R.D.;

12.15	*CLOSURE OF THE THIRD-DAY-SESSION*
12.20	Group 1*, bus departure for the excursion to the Radiotelescope ground station at Hooghalen (2 hours traveling by bus; a lunch-packet will be served during the tour).
12.45	Group 2* and 3* lunch in the Bastille-Restaurant.
14.15	Group 2, bus departure for the excursion to the International Institute for Aerial Survey and Earth Sciences (ITC) at Enschede.
	Group 3, bus departure for the excursion called "TWENTE TOUR", a sight-seeing tour through the surroundings.
20.00	*SYMPOSIUM DINNER AT THE ITC-GUESTHOUSE ENSCHEDE.*

* The groups will be composed during the symposium.

THURSDAY 12 APRIL 1979

SESSION 5 **INCANDESCENT LIGHT SOURCES; MATERIALS**

Chairman: Dr. L. Bartha, Hungarian Academy of Sciences.

9.00 *INVITED PAPER*
LAMP MATERIALS 93
Dr. J.R. Coaton, Thorn Lighting Ltd., Leicester LE4 7PD, U.K.

CONTRIBUTED PAPERS ON INCANDESCENT LIGHT SOURCES

9.45 36 ENERGY SAVINGS FOR AN INCANDESCENT LAMP FILAMENT IN A RADIATION REFLECTIVE SURROUND, 97
J. Brett, R. Fontana, J. Parascandola, W. Thouret, P. Walsh, S. Spura and L. Thorington, Duro Test Corporation, North Bergen, New Yersey, U.S.A.

10.00 37 SAG AND TILT OF INCANDESCENT COILS AT COMPARATIVELY HIGH TEMPERATURES, 99
H. Brons, W. van Hoorn and K.C. Thompson, Philips Lighting Division, Eindhoven, The Netherlands.

10.15 38 A FIELD-ION MICROSCOPE STUDY OF THE MICROSTRUCTURE OF TUNGSTEN FOLLOWING HELIUM ION IMPLANTATION, 100
J.M. Walls and R.J.K. Nicholson, Department of Physics, Loughborough University of Technology, Loughborough, Leicestershire, U.K.

CONTRIBUTED PAPERS ON MATERIALS

10.30 39 PASSIVATION LAYER OF ALUMINIUM FLUORIDE PROTECTS FUSED QUARTZ LAMP ENVELOPES FROM ATTACK BY FLUORINE, 101
J.R. Fitzpatrick and V.W. Goddard, Thorn Lighting Ltd., Research and Engineering Laboratories, Leicester LE4 7PD, U.K.

10.45 *COFFEE*

11.15 40 SINTERED YTTRIA FOR LAMP ENVELOPES, 103
W.H. Rhodes, GTE Laboratories, Inc., Waltham, Mass. 02154, U.S.A.

11.30 41 ANALYSIS OF HIGH PURITY METAL IODIDES FOR DISCHARGE LAMPS, 105
L. Bruninx-Poesen and G. Solleveld, Philips Lighting Division, Eindhoven, The Netherlands.

11.45 42 NEW GETTERING ALLOYS FOR DISCHARGE LAMPS, 106
E. Rabusin and M. Borghi, SEAS Getters S.p.A., 20151 Milan, Italy.

12.00 43 GASEOUS FLUORESCENCE AND VACUUM ULTRAVIOLET EXCITATION, 107
J. Maya, GTE Lighting Products Business, Danvers, MA 01923, U.S.A.

12.15 44 PROGRESS IN SAPPHIRE ENVELOPE LAMP FABRICATION AND APPLICATIONS, 109
L. Reed, N. Anderson, R. McRae and J. Richter, ILC Technology, Inc., Sunnyvale, California, U.S.A.

ADDITIONAL PAPERS:

12.30 ***CLOSURE OF THE SYMPOSIUM***
Prof.ir. H.G. de Winter, chairman of the organizing committee.

13.15 ***LUNCH AT THE BASTILLE-RESTAURANT.***

5 INFORMATION FOR SPEAKERS

5.1 SLIDES

Please hand your slides to the information desk, half an hour before the start of the morning or of the afternoon programme.

Please thumb-spot all slides in the lower left corner when the slides read correctly on hand viewing. Add sequence numbers.

5.2 INFORMATION FOR POSTER–SESSIONS

For those who haven the intention to supply material for the poster-sessions the following information is given:

1. The building is open at 08.00.
2. Materials will be provided attaching displays to the boards.
3. Please stay at your posters during coffee and tea breaks.
4. The posters will be accessible during the day of the presentation and if desired on the following day.

5.3 TIMEKEEPING

Speakers are asked to co-operate with the Chairman in keeping strictly to the allocated time (15 mins. per paper).

LIGHT-SOURCE TECHNOLOGY 1979 - A REVIEW

John F. Waymouth

A review of the present state-of-the-art of light source technology is developed in three stages: retrospective, overview of the present, outlook for the future.

I. Retrospective

A look backward into the history of electric light source development reveals the following facts:

A. The remarkably international nature of significant contributions.

B. A pattern of periodic revolutionary new light source developments followed by prolonged evolutionary further development, with life cycles of products measured in decades or generations. We can classify these stages as follows:

1. Revolutionary
2. Immature evolutionary
3. Mature evolutionary
4. Very mature evolutionary

C. The importance of empirical research and materials development in the foregoing sequence. Theoretical analysis seldom is a predictor of new light sources, and in fact usually does not approach conclusions until the mature evolutionary stage, as an identifier of the fundamental limits to performance. Moreover, the ultimate design and performance of the very mature evolutionary product is almost entirely dictated by the limitations of the materials of which it is made.

II. Overview of the Present

Within the framework of the above classifications we can locate the major types of general illumination light sources in the following categories:

A. Incandescent - very mature evolutionary
B. Fluorescent - very mature evolutionary
C. Low Pressure Sodium - very mature evolutionary
D. Mercury - very mature evolutionary
E. Tungsten Halogen - mature evolutionary
F. High Pressure Sodium - mature evolutionary
G. Metal Halide - mature evolutionary

Notable is the absence of a single revolutionary new type of light source. Based on a historical pattern, there should be one in the not-too-distant future.

III. Outlook for the Future

To attempt to foresee the possibilities of development of revolutionary new light sources, we ask two questions. First, where is the greatest need for new light sources? Second, where have we not already looked for new light-producing-mechanisms?

The answer to the first question is clearly in energy-efficient light sources for the home. However, it seems rather more likely that this need will be satisfied by revolutionary applications of fluorescent technology rather than by revolutionary new processes of producing light.

To the second question there are several answers. Radiation from molecules has been a significant factor only in tin halide lamps. Surely there must be other useful molecules, either as directly excited radiators or as fluorescent species. Second, although radiation from solids and gases has been extensively applied in light source technology, radiation from liquids has found application only in one-shot chemiluminescent devices. Surely there must be other ways of utilizing the liquid state in practical light sources.

IV. Summary

A review of the present state of light source technology suggests that the industry is presently in an evolutionary phase between revolutionary developments. No lamp developments that are not extensions of presently-known luminogenic processes are clearly evident on the horizon. Possible fruitful areas of research are identified.

The Design of High Pressure Discharge Light Sources

R. J. Zollweg
Westinghouse R&D Center
Pittsburgh, Pennsylvania 15235
U.S.A.

This review will be organized around four distinct but interrelated aspects of the design of high pressure discharge light sources: (1) the selection of the preferred radiators and radiation mechanisms, (2) the importance of the environment of the radiators, (3) the complications of getting the radiators into the discharge and their segregation there and (4) the arc structure or temperature profile and its relationship to arc constriction and instability.

(1) Although central to the design of high pressure light sources, the comparative selection of preferred radiators has received scant attention in the literature and much of that has been rather approximate. Part of this arises from proprietary considerations but much results from the complication of the problem. The relative photopic luminosity $(\bar{y})$, intended to represent normal eye response for a 2 degree field, is ordinarily used as the wavelength criterion in determining efficiency of possible light sources. However, a growing body of literature attests to the inadequacy of this criterion for some lighting applications and we should expect modifications and refinements in the future.

Radiation from excited atoms (Hg,Na,Tl,Ce) or molecules (SnI,SnCl,HgCl) plays a dominant role in most visible light sources. Continuum radiation resulting from the recombination of electrons and positive ions can be important in a few cases (Xe,Cs). Radiation associated with the attachment of an electron to a neutral atom can occur in the visible (O) and improve the efficiency of otherwise poor performers. Real arcs have a rather wide temperature distribution and several different radiation mechanisms contribute to the resultant light emitted. The channel model approximation can be used to estimate the performance of mercury and metal additive arcs.

(2) The major characteristics of high pressure discharges are successfully described on the assumption that the radiators are present in an environment in which local thermodynamic equilibrium (LTE) exists. However, non-LTE effects have been clearly demonstrated in several cases, usually involving spectral details. Saha equilibrium exists in LTE arcs and describes the equilibrium between neutral atoms, electrons and positive ions. This may have an important bearing on arc efficiency since the ion radiation tends to be predominantly of shorter wavelength and much less desirable than the neutral atom radiation. At the additive pressures used in high pressure discharges, self-absorption plays a very significant role especially for atomic radiators with a few strong lines. Although considerable progress has been made on this complex radiative transfer problem, more work is still needed. Buffer gases can

increase line broadening to reduce self-absorption as well as to increase the arc voltage and its time constant.

(3) At first sight the simple vaporization of atoms or molecules appears sufficient to transport the desired species into the arc discharge. However, many preferred additives have too low vapor pressures both as metals and as metal iodides. At the pressures encountered in high pressure discharges, a number of unexpected phenomena involving the formation of complex vapor molecules have been found which significantly increase the additive densities in the discharge. Molecular dissociation and the ionization of atoms combined with diffusion and convection result in both radial and axial segregation of additives in vertical high pressure discharges. Chemical reactions of additives and impurities with envelope materials and electrodes are important factors in the maintenance of lamp characteristics.

(4) The structure or temperature profile of a high pressure discharge can have an important bearing on lamp efficiency and arc stability. It is desirable to have a significant volume of the discharge at the optimum temperature and additive pressure since radiative efficiency can vary markedly with these parameters. If the arc becomes too constricted the stable cylindrical configuration may be transformed by magnetic forces to a helical configuration which usually rotates continually. Theoretical models have been successfully developed to predict the temperature profiles of vertical arcs in simple gases operated either on dc or ac circuits. Models of convection in both vertical and horizontal arcs have been developed. With some simplifying assumptions models have also been constructed of the axial segregation of additives that occurs in vertical metal halide lamps which results from diffusion and convection.

Accurate material transport properties needed for modeling of high pressure discharges were not available in the past but this situation has improved greatly in recent years. Some recommended sources are indicated below.

<u>Electron Drift Velocities and Cross Sections for</u>:

Hg - M. T. Elford, 31st Annual Gaseous Electronics Conf., p. 69, Buffalo, N.Y. Oct. 17-20, 1978 (abstract) and private communication.

Na,Tl,(Hg) - Y. Nakamura and J. Lucas, J. Physics <u>D11</u>, 325, 337 (1978).

Na - D. L. Moores and D. W. Norcross, J. Phys. <u>B5</u>, 1482 (1972).

<u>Atomic Transition Probabilities</u>

W. L. Wiese, et al., National Standard Reference Data Series NSRDS-NBS 4, Vol. I H-Ne, NSRDS-NBS 22 Na-Ca.

W. L. Wiese and J. R. Fuhr, J. Phys. & Chem. Ref. Data <u>4</u>, 263 (1975) Sc and Ti.

W. F. Meggers, C. H. Corliss and B. F. Scribner, Tables of Spectral-line Intensities, Parts I and II, NBS Monograph 145 (1975). Relative intensities of spectra with many weak lines including the rare earths.

SOME IMPORTANT FACTORS IN DESIGN OF A COLOR IMPROVED HIGH PRESSURE SODIUM LAMP.

By

Ranbir S. Bhalla
Westinghouse Electric Corporation
Lamp Division, Bloomfield, N. J. 07003, U.S.A.

Introduction: High pressure sodium (HPS) lamps have earned a well deserved reputation of a high efficacy, long life and a reliable source of light with good lumen maintenance. While the color rendition by the currently marketed HPS lamps is fair for many applications it is not very satisfactory for several other applications. To expand the use of this high efficacy source, for example in the offices and even merchandizing, it is essential that the color rendering properties be improved.

It has been known for quite some time that by increasing the pressure of the sodium vapor in the lamp arc tube the source color can be improved (1, 2). However, because of several technical problems associated with the design of a long life high color rendering lamp, a practical lamp has not been available in the market. This paper describes such a long life HPS lamp with improved color rendering (CRI of 65 as compared to about 20 for the standard lamp). Selection of materials to give a long lamp life will receive special attention.

Design Goals: The design goals included an improvement in the color rendering index (CRI) from about 20 for the standard lamps to about 65 for the new lamp. The loss in efficacy, as compared to the standard HPS lamps, should be less than 15%. It should be interchangeable with the standard similar wattage lamps (i.e. 100 volt operation for the 150 watt, 200 watt, 250 watt and 400 watt lamps). The goal for the lamp life was set at 7500 - 10,000 hrs.

Relationship Determination: In order to obtain an appropriate lamp design with the proper source color (CRI) the relationships between the electrical characteristics, sodium 'D' line reversal and CRI were determined for the arc tubes with the internal diameters of 5.4 mm, 7.2 mm and 8.0 mm. Sodium to mercury amalgam ratios (in % of sodium by weight) of 21% to 25% with 20 torr of xenon were used as the arc tube fill ingredients. These measurements showed that for an amalgam ratio of 25% a CRI of 65 is obtained for a sodium 'D' reversal width of 28 nm. This value of CRI is obtained at an approximate loss of about 10% in the lamp efficacy as compared to the currently marketed lamps which are designed for the optimum efficacy. Further details will be available elsewhere (3).

Materials Selection: Key to the design of color improved HPS lamps with long life was the selection and proper treatment of the various materials used in fabricating the lamps. New and improved electron emission materials had to be developed. Selection and treatment of the heat conserving shields at the arc tube ends was also found to be a very important factor in controlling the lamp voltage rise during the life. The most important factor in improving the lamp life was a big improvement in the lamp seal. The selection and the treatment of the materials will be covered in detail in the presentation of this paper.

References

(1) Schmidt, K., Proceedings of 6th international conference on Ionization Phenomena in Gases, Paris (1963)

(2) Mizund, H., Aicvtsu, H. and Watarai, Y., GTE XVII session, Barcelona, (1971)

(3) Bhalla, R.S., Larson, D.A., and Unglert, M.C., presented at the I.E.S. (U.S.A.) conference, Denver (1978). To be published in the Journal of I.E.S.

TOTAL EMITTANCE MEASUREMENT OF ARC TUBES IN HIGH PRESSURE SODIUM LAMPS

by E.F. Wyner, GTE Sylvania, 100 Endicott St., Danvers, MA 01923

A method has been developed to determine the total emittance of arc tubes used in high pressure sodium, HPS, lamps. Since the arc tubes in HPS lamps are contained within vacuum outer jackets, heat dissipation from the tube walls is primarily by thermal radiation, the magnitude of which is proportional both to the total emittance of the arc tube and to the fourth power of the wall temperature. To maintain a constant heat loss from the walls, the wall temperature will decrease as the total emittance increases. Wall temperature in turn affects many discharge parameters, such as absorption of sodium resonance radiation (D lines), reversal width of the D line, and the temperature at the amalgam reservoir, all of which affect the voltage and/or luminous output of the lamp. Total emittance is therefore a materials parameter which is inextricably linked to lamp operation and heretofore has not been measurable by any reported means sufficiently convenient for use by the lamp industry.

To determine emittance the temperature decay of the arc tube after extinguishing an operating lamp is needed. Temperature is determined from the intensity of the D line at 589 nm.[1] The method has been verified by use of a 7.9 μm infrared pyrometer for polychrystalline alumina (PCA) and sapphire arc tubes.[2] Over a temperature range from 1120°C to 1240°C, the agreement is within 10°C for alumina and 40°C for sapphire. The absolute value uncertainty is felt to be within 40°C for these measurements.

The total emittance is determined from an energy balance on the arc tube between the rate of loss of heat content and the radiated thermal power. Extensive studies in the late afterglow after extinguishing the arc indicate that from within the arc tube radiation is primarily in the $A'\Sigma - X'\Sigma$ sodium molecular band.[3] The atomic sodium resonance lines are in equilibrium with the walls[2] and their maximum intensities are, therefore, equal to that of a blackbody at the temperature of the wall. The magnitude of afterglow gas radiation is on a scale of one thousand to five thousand times less than from the operating arc. Since thermal radiated power is also on the same order of magnitude as arc radiated power, in the afterglow gas radiation can be neglected compared to thermal radiation from the wall as an energy loss mechanism.

The emittance as a function of wall temperature for a 400 Watt arc tube is shown in Fig. 1, Curve A. The emittance decreases as the wall temperature increases. The wall temperature at 400 Watts was 1162°C. The wall temperature and emittance values for this lamp have been compared to the values predicted from a computer model developed by Waymouth.[4] The data indicates agreement within 10°C between experiment and the model.

To illustrate the effect of varying arc tube emittance on wall temperature experimentally, the arc tube was remounted in an outer jacket, the inside surface of which was lightly coated with an infrared reflective coating. The effect of the coating was intended to reflect infrared radiation back to the arc tube, thus effectively reducing its total emittance. The effect in the afterglow is to reduce the rate of temperature decay. Curve B of the figure indicates a reduction of the total emittance versus temperature curve by approximately .015 with the coating. At 400 Watts the wall temperature increased 60°C with the coated outer jacket.

A first model to describe the total emittance of alumina tubes has been developed which divides the spectral emittance of alumina into two regions. The translucent region incorporates the visible and infrared spectral regions extending in the infrared to a transition wavelength beyond which the alumina becomes opaque(a spectral emittance of one). The emittance in the transparent region, according to the model, equals one minus the visible transmission of the arc tube. A calculated total emittance as a function of temperature curve is plotted as Curve C of Fig. 1 with a transition wavelength of 5.5 μm[5] and a spectral emittance in the translucent region of .05. An increase of 1% in visible transmission corresponds to approximately a decrease of .01 in total emittance. The sensitivity to the transition region is indicated by a decrease of .03 in total emissivity when the transition wavelength is increased from 5.5 μm to 6 μm. A decrease in emittance is observed with increasing temperature. However, the rate is less than that observed experimentally, Curves A and B. The reason for the discrepancy and observed variation in temperature dependence of arc tubes is under investigation.

References

1. J.J. de Groot, Comparison Between the Calculated and the Measured Radiance at the Center of the D-Lines in a High Pressure Sodium Vapor Discharge, Proc. 2nd IEE Conf. on Gas Discharges, London, P. 124 (1972).
2. E.F. Wyner, Bull.Am.Phys.Soc. 23, No. 2, P. 133 (1978), and GTE Sylvania Report LR-85, Mar. 1977.
3. E. F. Wyner and J. Maya, Bull. Am.Phy.Soc. 22, No. 2, p.198(1977) and GTE Sylvania Report LR-65, Jan. 1977.
4. J. F. Waymouth, J. of I.E.S., 6, 131 (1977).
5. Y.S. Touloukian, et al, Thermophysical Properties of Matter, Vol. 8 - Thermal Radiative Properties, IFI Plenum, New York-Washington (1972).

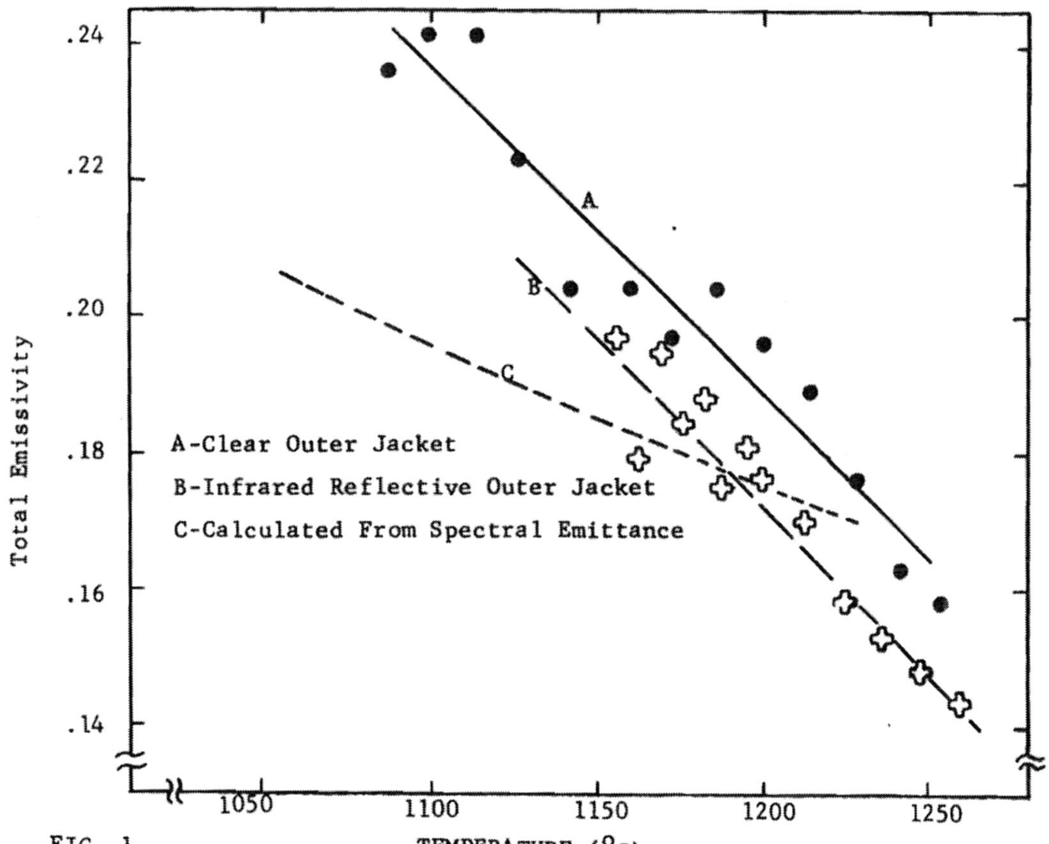

FIG. 1
Total Emissivity VS. Temperature for PCA 400 Watt Arc Tube.

DETERMINATION OF THE SODIUM AND MERCURY VAPOUR PRESSURE IN HIGH PRESSURE SODIUM LAMPS

J.J. de Groot and J.A.J.M. van Vliet
Philips Lighting Division
Eindhoven
The Netherlands

In high pressure sodium lamps the arc voltage is determined mainly by the partial vapour pressures of sodium and mercury which are controlled by the sodium-mercury ratio and the temperature of the amalgam[1].

Methods have been developed to determine

a) the sodium vapour pressure from the plasma temperature profile and the wavelength difference $\Delta\lambda_B$ between line centre and the maximum in the undisturbed blue wing of the self-reversed Na D-lines[2,3,4].

b) the mercury vapour pressure from the plasma temperature profile and the wavelength difference $\Delta\lambda_L$ between line centre and the maximum in the long wavelength wing of the self-reversed Hg 254 nm resonance line[4].

These methods are based on the assumption of resonance-broadening (Na-Na interaction) and van der Waals broadening (Hg-Hg interaction) for the resonance lines of Na and Hg respectively.

For practical high pressure sodium lamps, with mercury as a buffer gas and with a glass outer bulb, the above mentioned method of mercury vapour pressure determination cannot be applied as this outer bulb is not transparant for 254 nm radiation. In these lamps the ratio between the mercury and the sodium vapour pressure has been determined from the asymmetric contour of the self-reversed Na D-lines.

A numerical solution of the radiative transfer equation for the Na D-lines, assuming resonance broadening due to Na-Na interaction and van der Waals broadening due to Na-Hg interaction gives the dependence of the line contour on the sodium and mercury partial vapour pressures. From this it follows that in practice the ratio between the mercury vapour pressure p_{Hg} and the sodium vapour pressure p_{Na} can be derived from the ratio of the wavelength differences between line centre (589.3 nm) and the maxima in the blue and red wing (i.e. $\Delta\lambda_B$ and $\Delta\lambda_R$ respectively) with the help of the following equation :

$$\left(\frac{\Delta\lambda_R}{\Delta\lambda_B}\right)^2 = 1 + C\,(\Delta\lambda_R^{\frac{1}{2}} - 2.7)\,\frac{p_{Hg}}{p_{Na}}$$

where $\Delta\lambda$ is given in Å, and C is a constant.
The value of C has been determined from measurements on lamps with a known sodium and mercury vapour pressure (methods a and b); C = 0.018 for lamps with an internal diameter of 7.5 mm.

Using the spectroscopically determined values of partial vapour pressures as input data for the numerical solution of the energy balance equation, good agreement is found between the measured and calculated field strength and plasma temperature for lamps with various amalgam compositions i.e. with various mercury and sodium partial vapour pressures; this supports the correctness of the applied method.

References

1) C.A.J. Jacobs et al., J. IES (1978) 125
2) Teh-Sen-Jen et al., J. Quant. Spectros. Radiat. Transf. 9 (1969) 487
3) J.F. Waymouth, J. IES (1977) 131
4) J.J. de Groot and J.A.J.M. van Vliet, Bull. Amer. Phys. Soc. 23 (1978) 133

Summary of paper for 2nd International Symposium on Incoherent Light Sources at Enschede (April 1979)

Author: S.A.R.Rigden

A new approach to the design of Sodium/Scandium Halide discharge lamps for general lighting service

The two commonest high pressure mercury/halide discharge lamps used for general lighting service are the Sodium/Scandium and Sodium/Thallum/Indium combinations. The intrinsic arc efficacy of the former exceeds the latter but at the same time presents more serious technical problems in utilisation. These problems stem largely from the high chemical affinity of the additives for the structural components of the discharge tube, particularly the electrodes. This results in a unidirectional chemical change from the ideal composition with consequent colour change and losses in transmission of the arc tube. This paper describes a programme of work based on a horizontal burning 175 watt Sodium/Scandium lamp which has resulted in a lamp design having higher initial efficacy and lumen maintenance than has hitherto been claimed for lamps of comparable rating.

Early work was aimed at duplicating the results claimed by other workers, but without too much success, and it rapidly became evident that a successful design would be dependent on the solution of a number of material science problems associated with chemical stability between the lamp components and filling. It was established that initial efficacies as high as 110 lumens per watt were possible from a lamp of 175 watts rating, but the need for thorium as an electrode activator together with the use of caesium iodide as a discharge stabiliser, reduced this level by some 20%. Further losses in efficacy resulted from the arc bowing and being restrained by the upper wall of the arc tube, with consequent losses from thermal conduction. These problems have been overcome by optimising the wall loading to 16.0 watts/cm^2, offsetting the electrodes below the arc tube axis to allow the arc to float freely within the tube, and re-designing the electrode, using scandium oxide as an activator instead of thorium.

The well known effect of re-ignition spikes during run-up has been studied and it has been established that this effect is due to hydrogen. A relation between spike height in volts and percentage hydrogen by volume in the discharge tube has been obtained. Furthermore, it has been shown that large initial re-ignition spike levels usually result in poor lumen maintenance during life. Much can be gained in reducing the hydrogen content by preheating the iodides under a reverse flow of argon before commencing the exhaust schedule. Nominally water free materials give off moisture and elemental iodine when raised to red heat. The remaining hydrogen source (silica) can be effectively gettered in the arc tube by including a small addition of bromine as sodium bromide in the iodide additives. Further improvements in lumen maintenance and efficacy result from optimising the quantity and component ratio of the additives, it being preferable to bias towards a high sodium/scandium ratio consistent with the required source colour temperature not being adversely affected.

A simple modification of the seal assembly has resulted in a seal life improvement from 50% survivors at 5000 hours to no pre 5000 hour failures. The elimination of seal cracks finalised a viable lamp design with an initial (100 hour) efficacy of 100 lumens per watt, a 3000 hour efficacy of 80 lumens per watt and relatively small change during this period of colour co-ordinate or electrical characteristics.

LOW POWER, HIGH PRESSURE, METAL HALIDE DISCHARGE LAMP

K.E. BROWN and A.G. CHALMERS

THORN LIGHTING LIMITED

RESEARCH AND ENGINEERING LABORATORIES, LEICESTER, ENGLAND

The design of a compact discharge lamp suitable for interior lighting is demanding of the envelope materials and the chemical system contained therein. As well as being compact it needs to operate at low power, have good colour characteristics, have high efficacy and an extended life. The present state of knowledge suggests that metal halide discharges are the most likely to meet these exacting criteria.

In this paper, factors influencing the design of efficient low power metal halide discharge lamps will be discussed with emphasis on material properties, arc tube shape and heat conservation. By way of illustration, the construction and performance of a 100W tin-sodium halide lamp in a ceramic envelope will be described.

METAL HALIDE LAMPS WITH THE MOLECULAR RADIATION SPECTRUM

Sarytchev G.S., Naraikina G.A., Prikupets L.B., Gavrilkina G.N.
Moscow, U.S.S.R.

Qualitative analysis and quantitative estimates permit a large group of halogenide compounds to be distinguished, suitable for the development of light sources with the molecular radiation spectrum.
Radiant, spectral and electric discharge characteristics in the wide range of the possible basic parameters of the lamps are discussed. An effect of additives with a relatively low excitation resonance level on the halogenide molecular spectrum is analyzed. The principal components of the molecular spectrum are identified based on an experimental model with a test additive.
The results obtained and the future of such lamps used for various research and technological purposes are dealt with.

CHARACTERIZATION OF DIFFUSE AND CONSTRICTED HIGH PRESSURE ARCS BY MAGNETIC BALANCING

P.C. Drop and W.J. van den Hoek
Philips Lighting Division
Eindhoven
The Netherlands

A major problem encountered in the development of metal halide arc lamps is the constriction of the temperature profile. This phenomenon gives rise to a number of undesirable effects, in particular arc instability. The constriction is strongly related to the chemical composition of the arc plasma. To evaluate the influence of changes in the plasma composition and other arc parameters a quick and simple method is needed for measuring the degree of constriction.

Constriction can be studied by measuring the temperature profile. This method, however, requires rather involved procedures. Another method is the measurement of the size of the luminous channel of the arc by taking photographs. The results, however, are strongly dependent on the type of emission process and the absolute value of the temperatures. Consequently, this method is not appropriate for comparative studies. In the present paper we propose another method, to be used as a practical tool in lamp development.

In a horizontally burning arc the upward buoyancy forces can be balanced by a downward Lorentz force resulting from an appropriate external magnetic field. This results in a straight arc. In a simplified two temperature channel model it can be deduced that the buoyance force depends on an effective arc radius, the evaporated mass per unit length and both the core and mantle temperature. Since this force is rather insensitive to changes in the temperature values, for a balanced arc the applied magnetic force is a measure for the effective arc radius i.e. for the degree of constriction.

The method has been applied to different high pressure arcs. For Hg-arcs an approximately constant arc radius is found when the tube diameter, filling charge and input power are changed. For $SnI_2/SnCl_2/Hg$-arcs the arc radius is independent of input power and decreases with increasing amounts of tin halide.

For NaI/TlI/InI/Hg-arcs the arc radius decreases with increasing input power. For a mixture of rare earth iodide (RI_3) and alkali iodide the arc radius increases with increasing alkali/rare earth ratio in the salt filling. Increasing the input power for a RI_3/NaI/Hg-arc results in a strong constriction. A similar effect is found with RI_3/CsI/Hg-arcs, however, above a certain input power the arc becomes diffuse again. This last effect is in accordance with the mechanism proposed for the earlier reported magnetic field induced arc oscillations[1)].

1) W.J. van den Hoek and J.A. Visser :
Bull. Am. Phys. Soc. 23 (1978) 133

8.

A METHOD FOR DETERMINING THE TOTAL ACCUMULATED BURNING TIME OF LAMPS

Vincent D. Meyer, Paul L. Fortucci and Elliot F. Wyner
GTE Sylvania, 100 Endicott Street, Danvers, MA. 01923

Since the rate of permeation of helium through glass is exponentially dependent on temperature, the transport of helium from the atmosphere into lamps occurs in significant amounts only during lamp operation when the glass envelope is hot. Therefore, the amount of helium in a lamp at any particular time is directly related to the total accumulated time of operation. After many hours of operation the partial pressure of helium in a lamp approaches four microns, the partial pressure of helium in the atmosphere. The rate of permeation of helium through the glass envelope is dependent on the glass temperature, thickness and the type of glass. The helium accumulation can be predicted by calculations using the permeation constants for helium measured by Altemose[1].

One hundred and fifty watt Unalux high pressure sodium (HPS) lamps, which had been operated for known periods of time, were chosen to illustrate the method. The results reported here are for lamps returned after operation in normal lighting situations not for lamps operated under controlled conditions in a laboratory. The helium partial pressures, measured by mass spectrometry, show good agreement with a predicted curve which has been calculated by assuming an average outer jacket thickness, an average temperature during operation and the permeation constant for helium through 7720 glass.

To determine the effects of helium accumulation in the outer jacket of HPS lamps, the following measurements were made. The wattage, voltage, sodium D line width and relative luminous output of HPS lamps were monitored as the pressure of helium in the outer jacket of an operating lamp was increased under controlled conditions. The test results show that the helium pressure must exceed ten microns to produce a change of one percent in any of the spectral or electrical properties. Since the accumulated helium cannot exceed a pressure of four microns, it is concluded that the lamp performance is unchanged by the in-diffusion of atmospheric helium.

The method of determining accumulated burning times is well suited to HPS lamps and similar lamps since the outer jacket contains a getter and the getter removes most gases other than rare gases and methane. The method has also been applied to tungsten-halogen lamps which contain a high pressure fill gas.

The results produce a larger scatter of measured points around the curve calculated for fused silica and for an average thickness and operating temperature. The scatter is attributed to the difficulties in detecting several microns, or less, of helium in several atmospheres of fill gas (less than one part in 10^6) and the variations in operating temperatures.

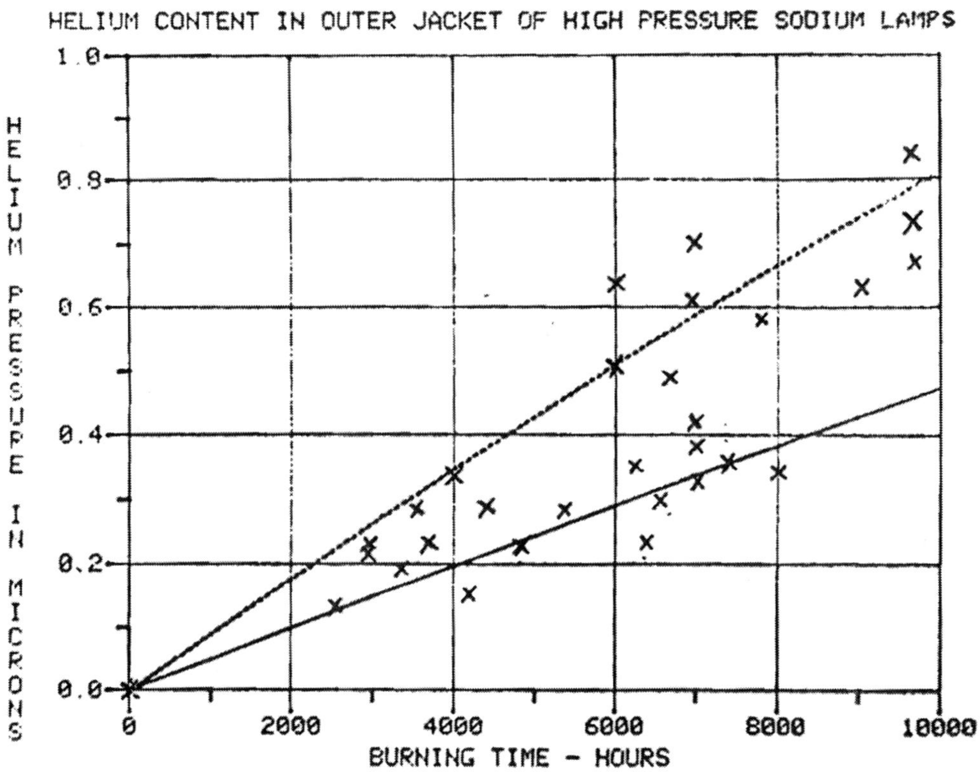

Dashed line - Lamp in enclosed fixture.
Solid line - Lamp in open fixture.
Two points at zero time are lamps only operated during processing.

[1] J. Altemose, J. Appl. Phys. 32, 1309 (1961).

HIGH PRESSURE ELECTRODELESS MICROWAVE DISCHARGES AS LIGHT SOURCES

by

Paul O. Haugsjaa, Robert J. Regan, William H. McNeill, Alfred E. Feuersanger
GTE Laboratories, Inc.
40 Sylvan Road
Waltham, Massachusetts 02154
U.S.A.

For many years, electrodeless discharges have been extremely useful as laboratory tools. They have served as spectroscopic light sources; sources of chemical free radicals; and atomic, molecular or ionic species in various states of excitation. Recently, however, electrodeless discharge technology has also become important to the lighting industry. Electrodeless discharges offer the advantages of electroded discharge light sources in efficiency and lifetime, without many of the complications that accompany the use of electrodes. Some of these complications include: tungsten migration via chemical equilibrium cycles, reactions of electrode materials with lamp components, and complex glass-to-metal seal design. Electrodeless discharges also offer some unexpected advantages. They retain high efficiency even when made very small, since the nearly constant power dissipation of electrodes is not required. They offer, in several cases, discharges which are rich in molecular emission, and they show excellent maintenance of light output over their lifetime (> 90% beyond 5000 hrs for ScI_3/NaI/Hg discharges).

W. Hittorf was the first to report, in 1884, the existence of a gaseous discharge without the use of electrodes.[1] The first U.S. patent on an electrodeless discharge was issued in 1926 to Leroy J. Buttolph and assigned to the General Electric Vapor Lamp Company of Hoboken, New Jersey. These early electrodeless discharges used induction coupling of RF power to the discharge from a coil placed around the discharge vessel. In this type of coupling, the coil/discharge assembly generally serves as a lossy component in an RF oscillator tank circuit. As the frequency of excitation increases, it becomes less practical to couple into the discharge by incorporation in a resonant circuit, so it is natural to use resonant microwave cavities to excite the discharge. One of the most popular microwave discharge cavities is the Evenson Cavity. This is cavity number 5 of F. C. Fehsenfeld, et. al.[2]

The termination fixture approach to excitation of electrodeless discharges is somewhat different. Since the discharge perturbs the resonant frequency of a cavity in a very significant way, it is possible to view the discharge itself as the electronic component into which microwave power is deposited. Thus, in what is called the Termination Fixture, the discharge serves as an electrical termination for microwave power traveling along a transmission line. The simplest termination fixture is just a coaxial line which has been expanded at one end to accept a discharge vessel. The discharge vessel rests between the center conductor and the screened-off end of the outer conductor, as in Figure 1. Incident RF power produces an electrical discharge in the vessel. Of course, it is unlikely that the impedance of the power source will match exactly the impedance of the discharge. Therefore, some significant fraction of the incident power will be reflected back to the generator. In order to match these impedances and insure efficient coupling of all the incident power to the discharge, a tuning element (such as a stub-stretcher) is required between the power source and the fixture. Though this scheme is perfectly acceptable for many laboratory applications, most commercial applications, particularly in lighting, would require the tuning to be accomplished within the termination fixture itself. This also reduces the losses associated with high-amplitude standing waves occurring over appreciable line lengths, and increases the practicality and efficiency of the device.

In order to accomplish this internal matching, one must first know the discharge impedance. The types of discharges discussed in this paper are of the class designated in the lighting industry as high pressure arc discharges. These discharges generally contain enough mercury to reach an operating pressure of several atmospheres. Our measurements show that a mercury discharge operating in a 700°C cylindrical envelope 1 cm long x 6 mm I.D. at 40 watts of 915 MHz power has an impedance of approximately 70 - j400 Ω. By carefully choosing dimensions of lines, ratios of diameters, and placing discreet impedances at proper locations, it is possible to match a 915 MHz discharge to a 50 Ω transmission line using only the space available in a termination fixture. The addition of other species such as metal halides to a mercury discharge normally changes the lamp impedance only slightly.

One quickly finds that matching impedances is only one of several key issues to be addressed. High pressure electrodeless discharges tend to become constricted near the envelope wall and attach themselves to the wall destructively, especially near the center conductor of the termination fixture where the applied power density is greatest. It is possible to eliminate this attachment process by modifying the lamp chemistry and the coupling geometry. For lamp fills which tend to be radiatively constricted (scandium halide fills, for example), the tendency for attachment can be reduced by the addition of a small amount of sodium halide. This addition tends to lower the axis temperature and broaden the radial and end temperature profiles, reducing the likelihood of arc constriction and wall attachment. It is of additional help to form the discharge vessel without sharp corners or narrow cavities and to place the lamp in a structure which minimizes RF fields at the wall where attachment normally occurs. These coupling structures may take the form of cups which provide a low field region at their center, or propagational structures, such as helices, which provide several low field-strength regions under large VSWR conditions.

When these matters of discharge chemistry, impedance matching, coupling geometry and arc isolation are addressed successfully, the result is a new type of incoherent light source. Figure 2 shows a plot of the luminous flux and efficacy of a 6 mm I.D. x 11 mm lamp with a $Hg/ScI_3/NaI/Ar$ fill operating in a 915 MHz termination fixture. At 50 watts of microwave power, this lamp has a cold spot temperature of about 660°C and a maximum envelope temperature of 750°C. At 50 watts of microwave power the light source produces about 95 lumens per watt of microwave power.

---

[1] W. Hittorf, Ann. Physik 21, 137 (1884).

[2] F. C. Fehsenfeld, K. M. Evenson and H. P. Broida, Rev. Sci. Inst. 36, 294 (1965).

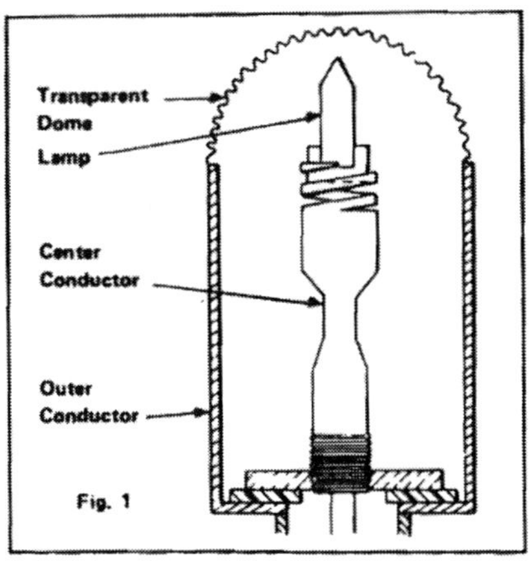

Fig. 1

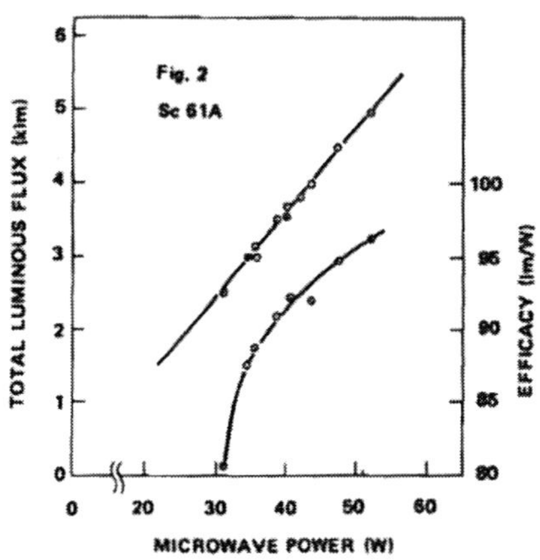

Synchrotron Radiation

(by Prof.dr. H.L. Hagedoorn, Eindhoven University of Technology, The Netherlands.)

High energy electrons circulating in an electron storage ring give rise to the emission of intense synchrotron radiation. This radiation covers a continuous wide spectral range that extends from 0.1 Å in the X-ray region to the infrared. The short wave length edge of the spectrum is characterized by the so-called critical wave length which is inversely proportional to the magnetic induction in the bending magnets of the ring and to the square of the electron energy.

The emitted radiation originates from a small source with dimensions of a few mm^2 to 10 mm^2 and is contained in a narrow forwardly directed solid angle of less than 1 mrad in the vertical plane and 1 mrad or more in the horizontal plane, depending on the dimension of that part of the electron orbit from which the radiation is considered. The total power of emitted radiation along the ring can amount to 100 kW or more.

This results in a number of emitted photons pro unit of solid angle and pro relative band width which is many orders of magnitude larger than in the case of classical photon sources. Due to this large intensity and the well defined nice geometry of the photon beam monochromators can be used with high efficiency. Therefore intense monochromatic, in wave length variable, photon beams can be delivered.

The spectral distribution of the synchrotron radiation can be accurately calculated and therefore used for calibration purposes.

The synchrotron radiation is used for many different applications:

- Solid state physics - surface research, photo-electron spectroscopy, crystallography.
- Atomic and molecular physics.
- Analytical chemistry - fluorescence techniques, absorption techniques.
- Physical chemistry - EXAFS, ESCA, catalysis, structure research, dynamic studies of enzymatic reactions.

- Molecular biology - structure research, fluorescense microscopy, contact microscopy.
- Medical physics - cell-structure research.
- Lithography - production of LSI circuits.

CALCULATION OF RADIATION ENERGY TRANSPORT IN OPTICALLY THICK HIGH PRESSURE DISCHARGES.

B.F.Jones *, D.A.J.Mottram *, and G.Eardley +.

* Thorn Lighting Limited, Research & Engineering Laboratories, Leicester.

+ School of Mathematics, Computing and Statistics, The Polytechnic, Leicester.

The radiant energy losses are difficult to calculate because either the data are unavailable or the required computations are too expensive. The problems are worse when the radiation cannot be approximated by the optically thin or thick limits, where the approach of Eardley, Jones, Mottram, and Wharmby (1978) for optically-thin many-line emitters or the diffusion approximation for optically thick lines can be adopted. For cases such as the sodium D-lines, where the line shapes and line broadening mechanisms are known as functions of the arc parameters, i.e. temperature, pressure and electron density, the radiation flux density can be calculated for a particular temperature profile by evaluating a quintuple integral (Church, Schlecht, Liebermann and Swanson, 1966). The radiation flux density is calculated at several radial positions, and it is therefore known as a function of temperature. The radiation flux density is substituted directly into the integral form of the energy balance equation; however, it must be treated as a function of temperature and not position. Alternatively, a set of net emission coefficients, which take account of absorption as well as emission, is calculated from the divergence of the radiation flux density. The net emission coefficients may be used in either the integral or the differential form of the energy balance equation.

Temperature profiles which have been calculated using radiation flux densities and net emission coefficients are compared in figure 1. The main advantage of using the radiation flux density is that it avoids the introduction of errors associated with the unnecessary differentiation of the radiation flux density to give net emission coefficients which are then integrated. The main disadvantage of this approach is that the radiation flux density is relevant for only one axial temperature, since it is zero on the axis for any cylindrically symmetric temperature profile.

A disadvantage associated with each method is that both the radiation flux density and the net emission coefficients are associated with a particular bore pressure and temperature profile. This means that when a temperature profile has been calculated, the radiation flux density and emission coefficients must be re-calculated, and the cycle repeated to convergence. This procedure is expensive in computer time.

The association of the net emission coefficients with a bore and temperature profile has been relaxed to a considerable extent by splitting the net emission coefficient into an emission component with an exponential temperature dependence, and an absorption component which is proportional to the radiation flux density at the volume element where the emission coefficient is being calculated:

$$\epsilon_N(T) = \frac{a}{T}\exp\left(-\frac{b}{T}\right) - \frac{c}{T}F_R \qquad (1)$$

where F_R is the accumulated radiation flux density, T the absolute temperature, and a,b, and c are constants which are found by fitting calculated net emission coefficients to this function.

The radiation flux densities for different temperature profiles calculated from the quintuple integral and using equation 1 show good agreement (figure 2).

References.

1. Eardley G., Jones B.F., Mottram, D.A.J., and Wharmby D.O., "A Model for a High Pressure Metal Halide Lamp Emitting Optically Thin Radiation", Colloquium on High Intensity Discharge Lamps, I.E.E., London, 11 May 1978.

2. Church, C.H., Schlecht, R.G., Liebermann, I., and Swanson B.W., A.I.A.A. Journal, 4, 1947-53, (1966)

Comparison of temperature profiles calculated using radiation flux density and net emission coefficients for a high pressure sodium lamp.

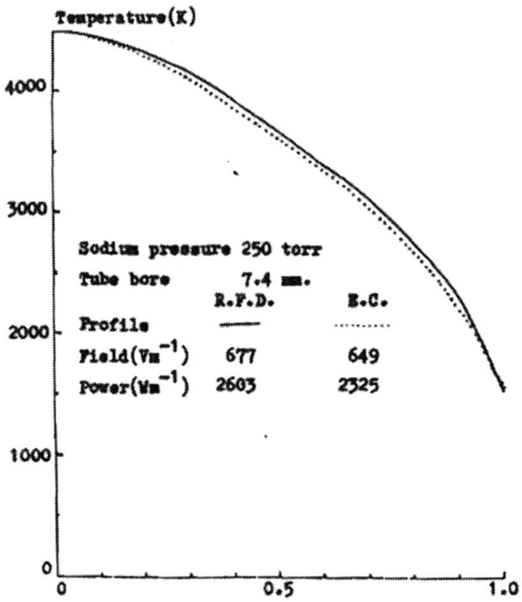

Figure 1

Radial variation of the radiation flux density for (A) a broad temperature profile, T_a=4000K,T_w=1200K, and (B) a parabolic profile, T_a=4500K,T_w=1200K.

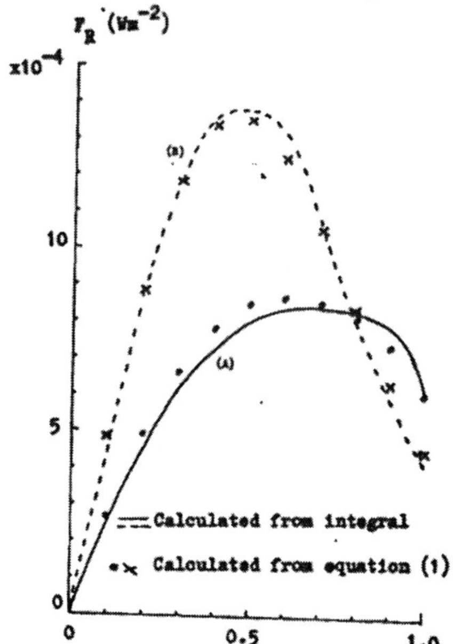

Figure 2

CALCULATIONS OF THERMAL CONDUCTIVITY AND OF THE EFFECTS OF RADIAL DIFFUSION ON ENERGY TRANSPORT IN HIGH PRESSURE DISCHARGES

D.A.J.Mottram and B.F.Jones.

Thorn Lighting Limited,Research & Engineering Laboratories,Leicester, England.

The solution to the energy balance equation gives information about the lamp temperature profile,spectrum and electrical parameters. All the discharges to be considered have a high pressure of mercury as a buffer gas; several additives will be examined, but particular attention will be paid to sodium because of its importance in discharge lamps.

The thermal conductivity of the vapour is calculated assuming that the inter-molecular forces which control the collisions can be described by a Lennard-Jones 6-12 potential. The two parameters required to describe this potential are the collision diameter,σ, and the potential depth,ϵ/k,(Hirschfelder et al). For mercury and sodium vapours, the σ are 2.898Å and 2.97Å respectively, and the ϵ/k are 851K and 1550K respectively. The calculations show that the thermal conductivity of a mixture of sodium and mercury does not deviate from the pure mercury value by more than 10% until the sodium pressure reaches half that of mercury. Consequently, the thermal conductivity of the vapour in the high pressure sodium lamp and most metal halide lamps is taken to be that of pure mercury.

Diffusive fluxes contribute to the energy transport. The contribution is proportional to the temperature gradient, and thus it is easily included as an additional term in the thermal conductivity.

The properties of a high pressure discharge lamp depend fundamentally on the chemical composition of the vapour. This composition may be calculated if LTE is assumed and if the relative abundances of the elements are known.

A simple approximation is to assume that the elemental abundances are constant at some estimated value (no demixing). In many systems, this approximation is in serious error due to radial diffusion. The concentration gradients estab-lished in the arc by the temperature gradients cause diffusive fluxes of all species. As the lighter species diffuse more rapidly,an imbalance is caused in the elemental abundances (radial demixing).

The elemental abundances resulting from this demixing have been calculated by solving the diffusion equation numerically for the steady state. Thermal diffusion and convection are neglected, ambipolar diffusion is included(McDaniel and Mason). The chemical equilibrium is found by minimising the Gibbs free energy of the system. In the case of radial diffusion, the composition is obtained as a function of temperature (Rogoff,1973) and not position.

The method has been applied to two types of discharge lamps: the high pressure sodium and the metal halide types.

In high pressure sodium lamps the only important reactions in the vapour phase are ionisation-recombination reactions. These cause demixing at high temperatures as the ambipolar effect increases the outward diffusion of sodium ions. Where the concentration gradients are highest, there is a peak in the thermal conductivity (figure 1).

In metal halide lamps, there are generally several elements present. The chemical reactions involved are thus more complex. However, some trends may be noted. The demixing effect is most marked where a light, multi-valent metal combines with a heavy halide, e.g. titanium and iodine (figure 2). It is less marked for a heavy metal, e.g. caesium, combining with the lighter halides.

The diffusive fluxes occur at lower temperatures, and hence one or more peaks may be seen in the thermal conductivity, particularly where the metallic additives are present in large amounts relative to the buffer.

References.

1. Hirschfelder, J.O., Curtiss, C.F., and Bird, R.B., "Molecular Theory of Gases and Liquids", Wiley: New York, 1964.
2. McDaniel, E.W., and Mason, E.A., "The Mobility and Diffusion of Ions in Gases", Wiley: New York, 1973.
3. Rogoff, G.L., Proc. 11th Conference on Phenomena in Ionised Gases, 206, 1973.

Acknowledgement : The minimisation subroutine was supplied by Dr R. Shawyer, N.P.L., Teddington.

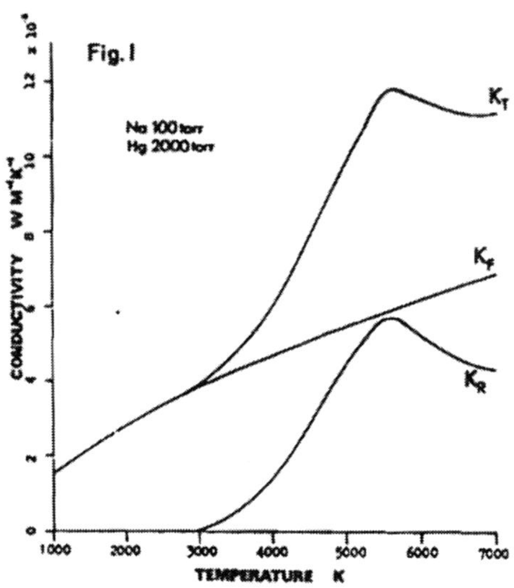

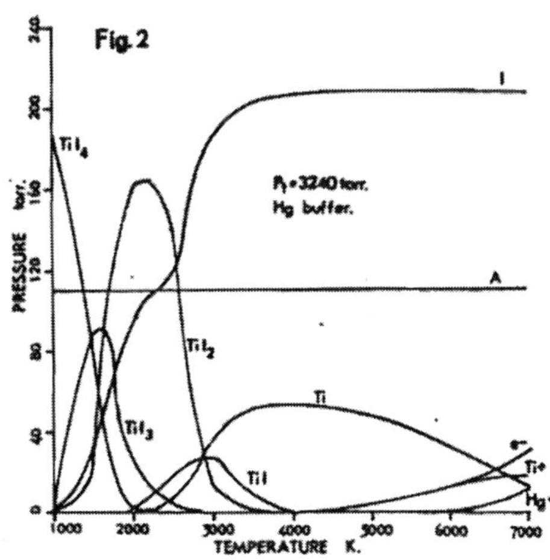

FREE CONVECTION AND QUARTZ WALL TEMPERATURES OF HORIZONTAL HIGH PRESSURE MERCURY ARCS

D. K. McLain and R. J. Zollweg
WESTINGHOUSE R&D CENTER, PITTSBURGH, PA., USA

When a mercury arc lamp is operated horizontally the arc rises due to the buoyancy of the hot gas until it is close enough to the quartz tube wall for the "wall stabilizing forces" to effect a balance. This asymmetry of the arc position causes the quartz wall temperature at the top to be higher than if the arc were centered. Two methods can be used to minimize the arc displacement: 1) an applied transverse magnetic field, or 2) a curved arc tube.

For an axially uniform arc we have developed a two-dimensional (r,θ) theoretical model which calculates numerically the temperature, velocity, and pressure distributions of the Hg gas and the temperature distribution of the quartz wall. This has been applied to mercury arcs at various pressures and power loadings, with transverse magnetic fields and with curved arc tubes. Most calculations of arcs in Hg have assumed the Boussinesq approximation, constant viscosity and constant density except for the buoyancy term, one exception being Lowke's model for vertical arcs.[1,2] For typical Hg-lamps the temperature varies from 1000°K to 6000°K and the denisty and viscosity change by factors of 7 and 4, respectively. We utilize the physical parameters of the gas, i.e., the density, viscosity, thermal conductivity, specific heat, emissivity, and electrical conductivity, as functions of temperature (the pressure changes only a few parts in 10^6 across the arc tube). The five equations solved are the r- and θ-components of the Navier-Stokes equation (momentum equation), the energy equation, the continuity equation, and Ohm's law. The equations are written in finite difference form and solved iteratively for the five unknowns, r- and θ-components of the velocity, temperature, pressure, and axial electric field. Heat is carried to the quartz wall by thermal conduction, by convection in the gas, and by optical absorption, and escapes by thermal radiative emission, by quartz conduction, and by conduction and convection in the gas of the outer, unless the outer is evacuated. The energy balance equation for the quartz wall, with the thermal conductivity and emissivity as functions of temperature, is also written in finite difference form and solved iteratively for the wall temperature.

The calculated locations of the arc core were found to agree with experimental observations to within about one mm, the approximate accuracy of the experimental observations. The calculated quartz wall temperatures were also in reasonable agreement with experiment.

Calculations show that the use of an applied transverse magnetic field changes the direction of the gas flow in the conducting region of the arc in agreement with early experiments of Kenty. Also shown is that a curved arc tube lowers the arc core through the mechanism of enhancement of the electric field in the lower relative to the upper part of the arc tube.

1. J. J. Lowke, to be published J. Appl.Phys. Nov. 1978.
2. R. J. Zollweg, J. Appl.Phys. 49 (3), 1077 (1978).

Effect of the free convection on the temperature distribution in a wall-stabilized horizontal arc.

T. Vicsek*

The temperature distribution in a wall stabilized horizontal arc is controlled by the heat conduction and by the gravitational field induced free convection. Starting from the energy, momentum and mass balance equations, the problem of the temperature distribution and flow field determination was solved using a finite difference scheme. Due to the results the central line of the arc is shifted upward, wich is consistent with the observations of F. Koury, W.C. Gungle, J.F. Waymouth: J. Ill. Eng. Soc. 4, /1975/, 106. The efficacy of a wall stabilized arc with free convection is compared to the efficacy of a convectionless arc.

*Institute for Technical Physics of the HAS
H-1325 Budapest, Hungary

DIAGNOSTIC STUDY OF A DyI_3/Hg - ARC PLASMA

W.J. van den Hoek and J.A. Visser
Philips Lighting Division
Eindhoven
The Netherlands

A study has been made of the radial distributions of Dy-atoms (Dy-I), Dy^+-ions (Dy-II) and iodine atoms in the core ($r/R \lesssim 0.5$) of a 50 Hz ac DyI_3/Hg-arc (Hg-pressure 4.5 atm., input power 500W, ac current 4.48 A rms, diameter 15 mm, electrode separation 3.7 mm). The arc was operated horizontally and magnetically centered.

The radial temperature profile of the arc was determined from the absolute side-on emission distribution of the optically thin Hg 577.0 nm line.

The Dy content of the arc was determined from the optical absorption of the Dy-II 495.7 nm and 445.0 nm resonance lines, in combination with calculations of the plasma composition taking radial depletion by ambipolar diffusion into account. The important role of ambipolar diffusion in establishing the radial distributions of Dy-I and Dy-II in the core of the arc follows from :

a) Measurements and calculations on the side-on emission distribution of the optically thin Dy-I 966.7 nm and the Dy-II 1052.3 nm lines, and
b) Measurements and calculations of the spectral radiances of the self-reversed non-resonant Dy-I 421.8 nm, 422.1 nm and 422.5 nm lines.

Due to the effect of ambipolar diffusion the total pressure of Dy-species at the arc axis is about a factor 10 lower than at the wall.

The iodine atom distribution in the core was evaluated from the emission of the optically thin iodine atom 905.8 nm line. The iodine atom pressure appears to be almost constant with radius. The value of this pressure is an order of magnitude higher than would be expected from the measured Dy-content. A reason for this discrepancy may be found in chemical reactions of DyI_3 with impurities (O_2, H_2O) or with the fused silica wall, which results in the formation of free iodine.

STUDY OF THE Na-D AND THE Tl 535 NM LINES EMITTED FROM A HIGH PRESSURE MERCURY DISCHARGE WITH SODIUM AND THALLIUM IODIDE AS ADDITIVES

H.-P. Stormberg
(Philips GmbH Forschungslaboratorium Aachen, Aachen)

We consider a cylindrical high pressure discharge lamp filled with Ar, Hg, NaI and TlI. The concentrations of NaI and TlI are determined by their partial pressures, which are controlled by the temperature and composition of the liquid iodide mixture at the coldest spot of the lamp. The calculation of the local plasma parameters of the steady state discharge has been performed using a measured temperature profile, assuming LTE and neglecting axial segregation. Then we can calculate step by step the following physical parameters:

a) the concentration-profiles of the species: Ar, Hg, HgI, HgI_2 , Hg^+, Na, Na^+, NaI, TlI, Tl, Tl^+, I_2 , I^+, I^-
b) the spectral absorption coefficients as a function of the radius
c) the spectral radiance of the lines.

A) The diffusion velocity of the species is mainly governed by the collisions with Hg-atoms because of the high concentration of Hg. Furthermore we have taken into account the ambipolar diffusion of the charged particles. The ambipolar diffusion can be calculated by introducing an additional electric field. All radial concentration profiles are then computated by an iteration procedure using the equation of continuity for each constituent.

B) Using these concentration profiles the local absorption coefficients for the Na and Tl lines are evaluated regarding mainly the broadening by Hg-atoms. Because of the high Hg concentration (compared to Na or Tl) Stark broadening or resonance broadening are not very important.

C) Using these absorption coefficients the radiation transport equation is solved numerically and the line shapes and the spectral radiance of the Na-D lines and the Tl 535 nm line have been obtained.

D) Additionally spectroscopic measurements on discharge lamps with different fillings of Hg, TlI, NaI have been done. The rather uncertain determination of the cold spot temperature can be substituted by a spectroscopic determination of the Na and Tl concentrations in the middle of the discharge. Using these concentrations the radial concentration profiles can be evaluated as well. Comparing now the calculated line shapes with the measured ones really good agreement is obtained. Furthermore the importance of diffusion and ambipolar diffusion becomes obvious. Neglecting these effects the agreement between theory and experiment appears to be rather poor.

SOME ASPECTS OF HEAT BALANCE IN METAL-HALIDE LAMPS

D.C. Fromm, J. Seehawer, W.-J. Wagner
OSRAM-Forschung, 8000 München 90, PF 90 06 20
Germany

The main objective of this work was to get a better insight into the power balance of high pressure metal-halide lamps in order to find out in which way the strong infrared losses of these light sources possibly could be reduced. The determination of the complete power balance is a difficult task because this kind of lamps usually consists of two main parts: The burner and the outer jacket. For each of both vessels a separate balance equation has to be fulfilled. As fare as we are aware a detailed analysis of metal-halide lamps has been published only once [1].

In order to enable the calculations two essential preconditions are nessessary. We have to know

1. the optical constants of fused quartz (for which material all our considerations are confined) as functions of wavelength and temperature
2. the heat transfer due to conduction and convection between the outer bulb and the ambient atmosphere.

So first of all we critically evaluated the literature data concerning these two points in order to provide most probable values of the complex refractive index of fused quartz and of heat transfer coefficients.

Object of our investigations was a 250 W Hg-Ar-Sn-Na-Tl-I-Br-lamp [2] with the following characteristic dimensions

1. Burner:
Electrode separation: 10 mm, volume: 1.1 cm^3, maximum inner diameter: 10 mm, wall thickness: 2 mm, total length: 75 mm.

2. Outer vessel (evacuated):
Inner diameter: 22 mm, wall thickness: 1 mm, total length: 130 mm.

On this lamp (operated in vertical position at 250 W, 50 Hz) we determined by infrared pyrometry the axial temperature distribution
1. at the outer vessel
2. at the burner, which was mounted in an outer vessel with a 2 mm axial slit through that the infrared measurements could be performed. Although such a procedure does not yield true burner-temperatures the result can be corrected by theoretical considerations.

Additionally side-on measurements were carried out in 5 spectral regions employing appropriate filter combinations covering without interruption a spectral range from 0.32 to 16 μ.

These investigations have been supported by decay-measurements after switching off the lamp, eliminating in this way the strong influence of arc radiation.

In order to determine the influence of the electrodes, their temperature distributions have been measured by optical pyrometry.

The preliminary results of our measurements and calculations are compiled in Fig.1. To some extent the specified values are based on rough approximations, mainly because of the lack of hemispherical measurements. Nevertheless the following unexpected conclusions seem to be justified:

1. The analyzed discharge as a whole is strongly radiation dominated.
2. The burner is heated by absorption of infrared arc radiation more than by convection and conduction. This conclusion is based on an estimation of the arc radiation in the region $\gtrsim 3.6\,\mu$.
3. A considerable amount of the infrared radiation of the burner penetrates through the outer vessel. Specially in the 3-4 μ region the outside detectable spectral radiance of the burner exceeds that of the outer vessel by far. This principally opens up the possibility of measuring the burner-temperature by infrared pyrometry without disturbing the thermodynamical system.

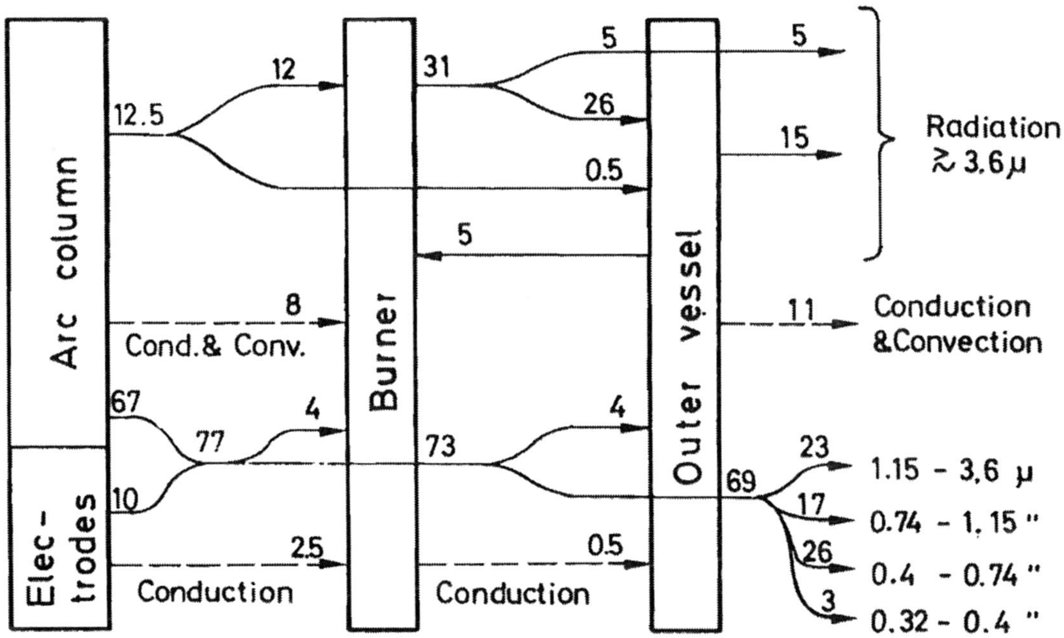

Fig.1. Power balance of a 250 W Hg-Ar-Sn-Na-Tl-I-Br-lamp (tentative). The quoted values are percentages of the total power.

Literature: [1] A.G. Jack, M. Koedam: J. of IES 3, 323 (74)

[2] D.C. Fromm, J. Seehawer, W.-J. Wagner: Light. Res. and Techn., in press.

CATAPHORESIS IN METAL HALIDE ARCS

H. Beyer, Akademie der Wissenschaften der DDR, Zentralinstitut für Elektronenphysik, IT. III, Berlin/DDR.

W. Funk, VEB Kombinat NARVA, Berlin/DDR.

H.G. Kloss, Akademie der Wissenschaften der DDR, Zentralinstitut für Elektronenphysik, IT. III, Berlin/DDR.

This paper is concerned with cataphoresis in metal halide arcs. Demixing processes arising from the electric field between the electrodes are able to change the concentration of the plasma components in the arc about several orders of magnitude. This effect has been investigated on arcs containing NaJ, TlJ and InJ additives. Longitudinal distributions of concentrations within the discharge tube have been measured. Calculations based on the impuls balances of plasma components also have been made and correspond to the measurements. It has been shown that cataphoresis is more effective for sodium compared to the other investigated additives.

EFFECT OF AXIAL SEGREGATION ON THE ELECTRIC FIELD STRENGTH IN 50 Hz Hg-Tl-I DISCHARGES

J. J. DAMELINCOURT, J. SALON, M. SEVIGNE - M. AUBES*
Centre de Physique Atomique, Laboratoire Associé au CNRS
Université Paul Sabatier, TOULOUSE, (France).

L. SCOARNEC et P. HERBET,
Compagnie des Lampes, Centre Technique,
7, Quai National, PUTEAUX (France)

This work studies three high pressure mercury lamps doped with thallium iodide. The discharges are operated verticaly on a 50 Hz supply with a choke ballast in series. The arc tube is a quartz tube with an inner diameter of 20 mm ; the distance between the electrodes is 100 mm. Other characteristics are as follow.

lamp	Mercury (mg)	Thallium iodide (mg)	Argon pressure at 28°C(torr)
T11	50	3. 3	20
T12	50	1. 6	20
T13	50	0. 8	20

The lamp wattage is 1000 w for the three discharges.

In order to treat the large amount of data due to Ac experimental conditions our measuring system has been conceived so as to work directly with a computer.

Temperatures and thallium atoms densities measurements.

By measuring the instantaneous and spatialy resolved spectral radiance of the unabsorbed lines of mercury and thallium we have determined temperatures profiles and particles densities at different heights.

Temperatures. In the calculation we supposed that atomic ratio mercury/argon is constant in a section of the discharge. We also supposed as a first approximation that the total pressure is constant in the whole discharge. With these assumptions an iterative process allows us to obtain, from measurements of 577 nm line intensities, temperatures values, mercury neutral atoms densities and pressures values, as function of time and position.

* Laboratoire de Physique des Milieux Ionisés, Université Paul Sabatier, TOULOUSE.

Densities of neutral thallium atoms. Densities of neutral thallium atoms are determined from measurements of 655 nm thallium line intensities assuming local thermal equilibrium.

We used the A_{ij} value from the NBS.

Calculation of the experimental atomic ratio and species densities.

Bearing in mind these results and by applying the classical laws of thermodynamics to our mixtures we obtain molecular and ionic concentrations. In these calculations the atomic ratio is assumed to be radialy constant and its value is chosen in order to fit the theoretical values of thallium densities in the central part of the discharge with the experimental ones.

Calculation of electrical condutivities and electric field strength.

From the precedent results it is then possible, with the help of the classical Chapman-Enskog theory to determine the profiles of electrical conductivities and from Ohm's law and current intensities to calculate the electric field strength.

Results.

Measured temperatures are given versus time for the three lamps and versus time and radial densities for the T11 lamp, respectively at 2 cm, 5 cm and 8 cm from the lower electrode.

The experimental and calculated atom thallium densities are given for the T11 lamp versus time, radial distance and axial position. The variation of the axial atomic ratio is given for the three lamps versus axial position.

As a test of validity the calculated values of the instantaneous electric field strength are compared to the values deduced from volt-ampere characteristics and electrodes fall measurements. From the results concerning the electric field we can conclude that measured densities and temperatures combined with thermodynamical model give a reasonably good description of the Hg-Tl-I discharge. Furthermore this method allows us to show the non uniformity of the electric field as a function of the height.

COMPUTATION OF RELATIVE VARIATION OF NET EMISSION COEFFICIENT IN TWO Hg-Tl-I DISCHARGES

J. J DAMELINCOURT, J. SALON,
Centre de Physique Atomique, Laboratoire associé au CNRS
Université Paul Sabatier, TOULOUSE (France)

L. SCOARNEC, P. HERBET,
Compagnie des Lampes, Centre Technique,
7, Quai National, PUTEAUX (France)

The work presented here is a first attempt to investigate the possible effect of A. C. supply on the radiation trapping The principle of the calculation is as follows: energy balance equation together with Ohm's law are solved numerically, for ten instants of the A. C cycle, using as boundary values the experimental values of central and wall temperatures ; in the energy balance equation the net emission coefficient is written under the form

$$\varepsilon_N(T) = \alpha\, U(T)$$

Where U(T) is a time averaged value of the net emission coefficient and α is modified, each time, in order to fit the theoretical value of the instantaneous arc current with the experimental one.

With this procedure, we derived values of α and electric field strengths as function of the time.

Theoretical model.

In order to solve the energy balance equation the following assumptions are made :

- the arc is radialy symmetrical,
- the hypothesis of partial thermodynamic equilibrium is valid in the whole discharge,
- the atomic ratio Hg/Tl is radialy constant,
- the diffusion fluxes are neglected.

For each section of the discharge we can write at time t :

(1) the energy balance equation : $\sigma E^2(z,T) = U(z,T) + \mathrm{div}(-\lambda(z,T)\,\mathrm{grad}\,T)$

(2) the Ohm's law : $I = 2\pi R E(z,T) \int_0^1 \sigma(z,T,\rho)\,\rho\, d\rho$.

In these equations, σ is the electrical conductivity, U the net emission coefficient of radiation, λ the total thermal conductivity, σ , λ depend upon atomic ratio R_a (for Hg/Tl) and temperatures ; they are calculated as function of T for each experimental value of R_a.

For the net emission coefficient of radiation, we used as starting values the ones proposed by Zollweg, Lowke and Liebermann (1) for mercury-iodine discharges.

Results.

Calculated values of the electric field strength are in reasonable accordance with experimental ones for lamps T13 which contains few thallium atoms ($2000 < R_a < 8200$) ; for lamp T11 ($70 < R_a < 350$) discrepancies are more important (20 to 40 per cent) but the opportunity of the actual choice of U for this lamp is doubtful.

For the region of zero current, the approximate steady state equation (1) is not very adequate and some disagreements are to be noticed between theoretical predictions and experimental results

Calculated values of α show that at low temperature U would be larger and smaller during the current maxima. These variations are possibly due to the diffusion of the energy toward the wall during the low current period.

(1) R. J. ZOLLWEG, J J. LOWKE, R. W. LIEBERMANN : J. Appl. Phys. Vol. 46, n° 9, p. 3828-3838 (1975).

LOW PRESSURE DISCHARGES

A.G. Jack
Philips Lighting Division
Eindhoven
The Netherlands

The low pressure discharge forms the basis of two important families of lamps, namely the fluorescent lamp based on the low pressure mercury-rare gas discharge and the low pressure sodium discharge lamp. Such discharges, which form the basis of efficient lamps, have typically a metal vapour pressure in the millitorr range and a rare gas pressure in the torr range. In these discharges radiation transfer plays a dominant role since typically 40 to 75% of the input power is converted into radiation.

Measurement of the plasma parameters can give insight into the processes which take place in the discharge and enable a detailed analysis of the discharge to be carried out. Various experimental techniques give information about the electric field, electron densities and temperature, the concentrations of some of the excited states and the total radiation output of the discharge. For the low pressure mercury-rare gas discharge a detailed analysis is available and numerical modeling of the discharge has been carried out. Because of the dominant role played by radiation in such a discharge this must be treated in some detail in the numerical models. The most recent extensions to the modeling of the discharge have been the introduction of a two electron group model rather than a single Maxwellian distribution and the inclusion of excited state collisional ionization. This has led to an improved agreement between the numerical models and the various experimental measurements.

All this gives a detailed description of the discharge of a fluorescent lamp. At present there is a search for fluorescent lamps with still higher luminous efficacies and also more compact lamps with higher loadings. With the development of narrow emission band phosphors high efficacy lamps with excellent colour rendering properties have been realized. The efficacy has been still further increased by decreasing the tube diameter from 38 to 26 mm in order to optimize the combination of discharge and new phosphors. Higher lamp efficacies can also be realized by operating not at the mains supply frequency but at higher frequencies, e.g. in the kilohertz range. At still higher frequencies electrodeless discharges become feasable and may allow more freedom in the choice of the geometry of the discharge. Various systems with operating frequencies varying from the kilohertz to the gigahertz range are described in the literature.

For highly loaded discharges the increase in the current density leads to a less efficient discharge. One way of minimizing this has recently been proposed : the introduction of a recombination structure into the discharge in order to increase the ionization loss and thus increase the electric field. For all discharges where the loading is increased there is an increase in the wall temperature and thus the mercury vapour pressure increases. In order to keep the mercury vapour pressure at its optimum value use can be made of an amalgam.

The low pressure sodium discharge is in many ways similar to the mercury discharge. However there are three differences, namely the need for a sodium resistant glass, thermal insulation of the discharge tube and the occurrence of depletion of sodium in the discharge axis. Because the optimum sodium vapour pressure of a few millitorr is reached at a temperature of about 260°C the thermal losses from the discharge have to be minimized. There have been a succession of improvements in the thermal insulation leading to the present day indium oxide filter. Because of the relatively high current densities in the sodium discharge there is often a high degree of ionization of the sodium atoms. Radial depletion of the ground state arises due to the relatively rapid ambipolar diffusion of sodium ions to the wall where recombination takes place and the slow back diffusion of neutral atoms into the discharge space.

As with the mercury discharge there have been extensive theoretical and experimental studies of the positive column of the low pressure sodium-rare gas discharge. As far as the numerical modeling of the discharge is concerned recently it has been shown that multistep electron impact ionization processes are important.

Recent improvements in lamp performance have been obtained by minimizing the migration of sodium due to thermal gradients and thus realizing a better control of the lamp voltage during the life of the lamp. Studies have also been made of the effect of modifying the discharge current waveform. As the current waveform is altered from a sinusoidal one to a more square waveform the column efficiency increases while the reignition voltage at each half cycle decreases due to the more rapid current reversal. A new lamp ballast has been introduced which makes use of these advantages.

For the low pressure sodium lamp, in the positive column about 43% of the electrical energy is converted into radiation. This is lower than the 74% typically found in the positive column of a fluorescent lamp. However after accounting for the conversion loss in the fluorescent powder and the electrode losses, the fluorescent lamp only converts about 25% of the input electrical energy into visible radiation. For the low pressure sodium lamp no phosphor layer is necessary and about 35% of the input electrical energy is converted into visible radiation.

Homogeneous recombination structures in depleted discharges (low-pressure sodium)

J. Hasker
Philips Research Laboratories, Eindhoven, The Netherlands

The influence of a recombination structure (consisting of glass wool being thinly and homogeneously distributed within the discharge space) on mercury-noble gas discharges may be well known[1,2]. As an application, the use of such a structure allows much shorter fluorescent lamps while maintaining high overall luminous efficacy at fixed lamp power and lamp current.

The results referred to above apply to discharges with negligible depletion of the (mercury) ground state. Low-pressure sodium (l.p.s.) lamps, however, show substantial radial depletion[3]. Therefore, the examinations on the influence of a recombination structure have been extended to this type of discharge.

Our experiments were performed on the well-known "35 W" l.p.s. lamp. The results to be mentioned apply to optimum sodium vapour pressure. The structure density was varied by applying various structure weights. Because lamp properties depend on the sodium distribution within the discharge tube[4] and on the processing, a number of reference lamps without structure have been made in the same way as the structure lamps. The lamp efficacy versus the structure density shows a maximum. At this maximum, the lamp efficacy, the axial electric field and the lamp power have been increased by about 10%, 40% and 5%, respectively, when compared with the reference lamps. This means that, for operation an optimum ballast circuitry[5], the increase of the overall efficacy will amount to about 20%.

In the absence of depletion the structure-inherent trends in the axial electric field, the electron temperature and the discharge contraction can be understood with the aid of a Schottky-type model for the positive column[2]. In the case of depletion the use of this model is more complicated because then the axial electric field does not only depend explicitely on the ionization rate but also on the electron and ground-state density profiles. Hence, both the continuity equation for the electrons and the ground-state atoms has to be solved. This was done for the usual boundary conditions[3] and under the assumption that the number of ions and electrons collected by the structure per unit volume and per unit time is equal to the number of ground-state atoms released by the structure. The structure-inherent increase of the electron temperature is taken into account in the iteration procedure by means of an one-step ionization process.

Calculation of the electron and ground-state density profiles shows that - at the above-mentioned maximum - the discharge with structure has less depletion than the conventional discharge while the calculated increase of the electron temperature is about 6%. It is noted that also in the absence of depletion the lamp efficacy versus the structure density shows a maximum[2]. This suggests that the structure-inherent increase of the electron temperature is important to this behaviour. As in the case of negligible depletion, the structure gives rise to contraction of the electron density profile at moderate depletion. However, at substantial depletion the opposite effect occurs.

1) J. Hasker, Appl. Phys. Letters, 28, 586, 1976.
2) J. Hasker, Journal I.E.S., 6, 29, 1976.
3) H. van Tongeren, Philips Res. Repts, Supplements 3, 1975.
4) M.H.A. van de Weyer, Philips Techn. Rev., 23, 246, 1961.
5) H.A.G.S. Smulders, Lighting Res. and Technology, 9, 65, 1976.

PHYSICAL PHENOMENA ON ELECTRODES AND NEAR ELECTRODE REGIONS IN LOW PRESSURE GAS DISCHARGE LAMPS

E.V. Okhonskaya, Mordovian State University named after N.P. Ogarev
Saransk, U.S.S.R.

The character of electrical conductivity changes, thermoemissive current, the temperature of electrodes in the low pressure mercury vapour gas during the operation of the lamp on the alternating and direct currents with and without electrical field near the cathode surface have been investigated using electrophysical methods.
The parameters of the near cathode regions have been determined by the help of the spectral methods. It was shown that the diffusion of alkaline earth metals atoms as well as the return of these of atoms on the cathode influence greatly on lamp life and electrode characteristics.
The theoretical method of calculation of the electrode electrical regimes during operation is reported.
The author also presents the experimental method of lamp life prognosis for low pressure gas discharge lamps.

A Circuit for Dimming Fluorescent Lamps with High and Low Resistance Cathodes

D.F.Chapman, Thorn Lighting Ltd., Research and Engr. Laboratories, Enfield.

The cathodes in a fluorescent lamp have to provide satisfactory electron emission during starting and operation. If the cathodes are operated too cold, much of the electron emission is obtained by ion bombardment and this gives a high cathode drop, increases cathode sputtering, forms heavy discolouration or blackening and reduces lamp life. If the cathode is operated too hot, excessive evaporation of the barium emitter takes place, discolouration in the form of end spots will appear and again lamp life is shortened. The optimum temperature of the cathode "hot-spot" during operation is about 900 - 1000°C.

Cathode types

Fluorescent lamps were originally designed for use in starter switch circuits in which the preheat current for starting was controlled by the impedance of the ballast. During operation in this circuit the cathode is "self-heated" by electron and ion bombardment and by the power loss in the tungsten wire caused by the discharge arc current flowing through part of the cathode coil. The tungsten wire cathode coils were therefore made long with High Resistance (HR) in order to obtain good life. In 1952 in the U.S.A., a new "Rapid Start" circuit was introduced to eliminate the starter switch and this used transformer windings in the ballast to supply a constant voltage to each lamp cathode during starting and operation. To reduce the power loss in the electrodes new Rapid-Start lamps with 3.6 volt cathodes having Low Resistance (LR) were introduced.

At the present time, I.E.C.81 (Recommendations for tubular fluorescent lamps) specifies 20,30,40,65,80 watt lamps with both HR and LR cathodes and in many countries both types are available. The lamps are interchangeable and the two different cathode types are not easily identified by the user. Problems can arise if the ballast and circuit is not matched to the correct type of cathode and lamp. For a given voltage the LR type is always hottest.(Fig.1.)

Dimming Circuits

Dimming is obtained by reducing the arc current. Present day systems use phase-control circuits with thyristors or triacs to vary the conducting period of each half cycle of arc current (Fig.2). Dimmers are available to handle up to 2½ KW of fluorescent lighting and control may be manual or automatic. Facilities such as auto-fade, pre-set levels, over-ride, colour change and photo-cell control can be included. On 220-240V supply, lamps up to 1800mm length can be dimmed. To obtain good stable dimming over a full range, special lamp ballast circuits are required, together with precise "gate" control for the semi-conductors.

At low levels of arc current there is insufficient electron bombardment during the anode half cycle and insufficient ion bombardment during the cathode half cycle to provide "self-heating" of the cathodes and it is necessary to provide continuous heating from an external source. Until recently, cathode heating circuits were "Voltage-fed" from preheat transformers giving either 3.6 to 4.0 volt or 6.3 to 7.0 volt output to each cathode. To ensure satisfactory operation and lamp life, it has been necessary to match the preheat transformer to the type of cathode. For example, operation of a HR cathode on a 3.6 volt transformer will result in low cathode temperature, difficult starting and possible tube end blackening. On the other hand, operation of a LR cathode on a 7.0 volt transformer will result in too high temperature, short life, end discolouration and possible overheating of the transformer.

New Cathode Heating Circuit

In phase-controlled dimming circuits the power-factor correction capacitors cannot be connected across the load because the voltage waveform from the dimmer output has a "chopped" shape and is non-sinusoidal. The capacitors must be connected across the supply input to each lamp circuit and the power factor obtained varies depending upon the dimming level. If the primary winding of the cathode heating transformer is connected in series with the capacitor(Fig.3) the current remains almost constant when the secondary

windings are loaded with either type of cathode. The cathodes are thus "current fed" and their voltage adjusts automatically to about 4.0 or 7.0 volts for either LR or HR cathodes.
The value of each power-factor correction capacitor varies with the lamp size and is chosen so that input current in the dimmed position is nearly equal to that in the full-up position. One transformer is used for lamp sizes between 600 and 1800mm and is designed so that magnetic flux saturation limits the series resonance which occurs when the cathodes are off-loaded(e.g.a lamp is removed).
In circuits with continuous cathode heating, the temperature of the cathode coil and lead wires is affected by the magnitude and phase of the heating current and arc current. The phasing of heating current from the transformer enables arc current to partly oppose circulating cathode current (Fig.4) A feature of the circuit is that power wasted in heating the cathodes is reduced by about half in the full-up dimmer position. For good dimming it is usually necessary to provide additional re-ignition aids in the form of earthed metal strips near the surface of the lamp tube and to generate ignition pulse voltages at the commencement of each half cycle of lamp arc current.
To achieve good dimming, the controls, dimmers, lamps, ballasts and luminaires should be compatible and designed and installed as a complete system. On stable supply mains with suitable solid-state dimmers the new circuit enables 40W and 65W fluorescent lamps to be dimmed down to a minimum 100:1 lumen ratio.

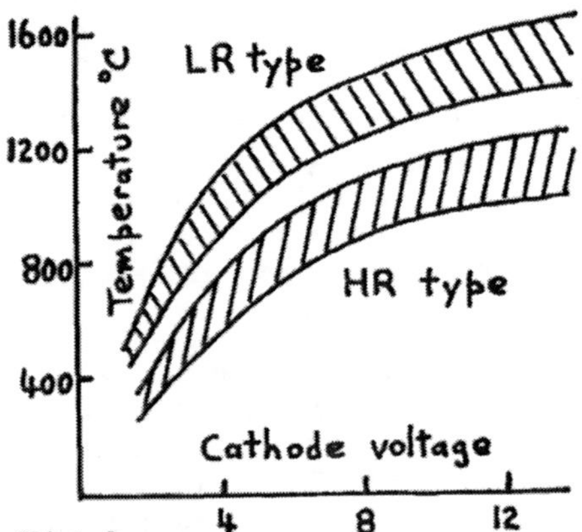

FIG.1.
40 W. Cathode characteristic.

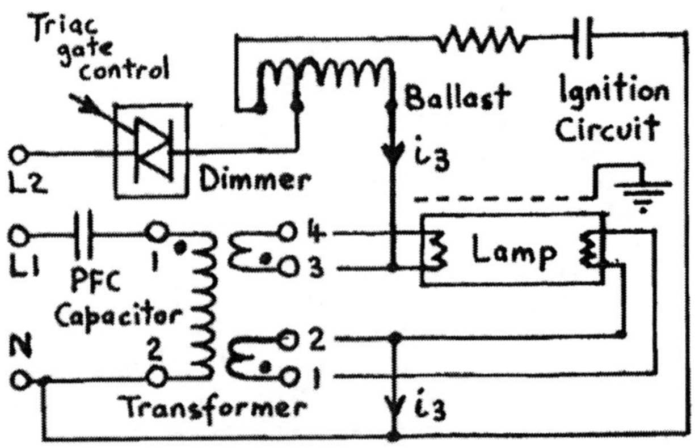

FIG.3. New fluorescent lamp dimming circuit using current-fed cathode heating for HR or LR cathodes.

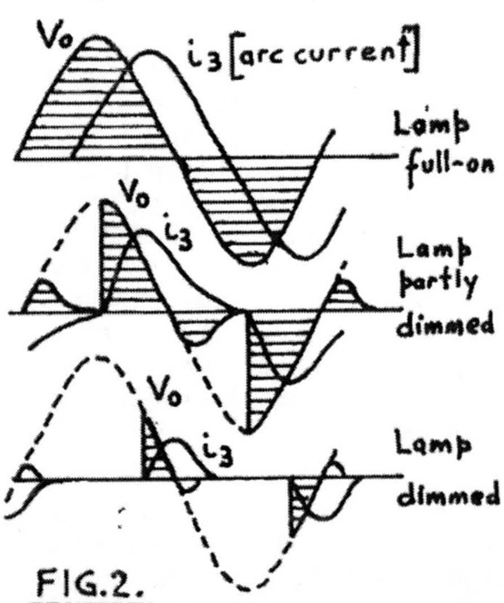

FIG.2.
Waveforms of dimmer output voltage and lamp arc current.

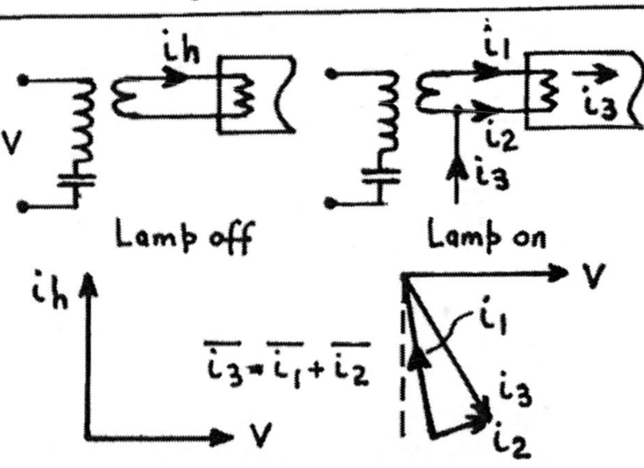

FIG.4. Phasor diagram for circulating currents in cathode leads. When the lamp is fully-on the cathode heating current reduces from i_h to i_2.

THE PHOTOMETRY OF INCANDESCENT LIGHT-SOURCES; THE RECENT DEVELOPMENTS AND THE CHANGES TO THE PHOTOMETRIC SI-UNITS IN THE NEAR FUTURE.

The Thirteenth General Conference on Weights and Measures in October 1967 adopted the following definition of the unit of luminous intensity, the Candela (cd):

"The candela is the luminous intensity in the perpendicular direction of a surface of 1/600 000 m^2 of a black body at the temperature of freezing platinum under a pressure of 101 325 Nm^{-2}.

It is a pity that the unit candela has been introduced into the International System of Units as a base unit. Jurgen R. Meyer (1) in 1968 and J.S. Preston(2) in 1969 discussed this matter of redudancy and illocicallity.

The main disadvantages of the present system are

1. The practical realization of the primary standard by means of a black body furnace is expensive, inconvenient and less reproducible than is desired.

The difficulties arise from

1a. The uncertainty as to the value of the temperature T of freezing platinum on the thermodynamic temperature scale.

1b. Other experimental difficulties as evidence of systematic differences between the measurements of different national laboratories.

2. The historical choice of intensity as a basic quantity is now seen to have been unfortunate.
   Intensity is a fairly complicated concept less readily understood than flux.
   It would be preferable if the system of photometric concepts could be presented in a similar way to radiometry, which is based on the unit of power the *watt.*

Definition of photometric quantities based on V(λ); recent developments.

With V(λ) as the normalized photopic eye-sensitivity as a function of wave length and $I_e(\lambda)$ as the radiant power proceeding from a point-source ($\frac{1}{6}.10^{-6}\,m^2$) of a black body at approximately 2042 K *) one calculated the value of the integral

$$I_v\,(T = 2042\,K) = \int_0^\infty I_e(\lambda, T = 2042\,K)\,.\,V(\lambda).d\lambda$$

and one found 0.00147 Watts sr^{-1}
also named light-Watts sr^{-1}.

Calculation, referring to the definition of the candela gives:

0,00147 light-Watts sr^{-1} = 1 candela
and 1 light-Watt sr^{-1} = 680 candela

This constant is called the maximum luminous efficacy of radiation Km; it relates the lumen to the Watt.

* Quinn: 2040.8 ± 0.3 K

Between the years 1975 and 1977 a number of (nine) national laboratories reported recent experimental values ranging from 675 to 686, with the UK National Physical Laboratory and four others opting for 683 $lm\ W^{-1}$

Thus in 1977 the value Km = 683 lmW^{-1} was recommended for the spectral luminous efficacy of monochromatic radiation of frequency 540.0154×10^{12} Hz.

Furthermore it was recommended in 1977 that the lumen be defined as follows: "The lumen is the luminous flux of a monochromatic radiation whose radiant flux is $\frac{1}{683}$ W and whose frequency is $540.0154 = 10^{12}$ Hz."

In 1977 Dr. J. Terrien, Director of the International Bureau of Weights and Measures proposed that the value of Km = 1 should be considered; this proposal was one of the most challenging contributions of the discussion. The use of a dimensionless Km would mean that all photometric measurements would be expressed in terms of watts.

Conclusions

1. The value Km = 683 $lm\ W^{-1}$, approved by the CIPM, should be adopted for immediate use. This value is consistent both with the current theoretical estimate and with the practical photometric scales maintained by national laboratories.

2. It would be preferable to replace the candela by the lumen as the photometric base unit of the SI immediately.

3. In the far future the photometric units can be abandoned by setting Km = 1, without dimensions.

4. The absolute accuracy of measurement of 'detector' spectral quantities may improve considerably as a result of the changes; measurement of 'source' spectral quantities may not be improved.

5. There will be no change in situation regarding any systematic discrepancies that may exist in practice between the various source-based and detector-based optical radiation scales.

Literature

1. Holst, G, Electrische lichtbronnen en hare eigenschappen, F. Behn, Haarlem, 1920.
2. Keitz, M.A.E., Light Calculations and Measurements, Macmillan, 1971 (2nd ed.).
3. J.R. Meyer-Arendt (1968), Radiometry and Photometry: Units and Conversion Factors, Applied Optics Vol. 7, No. 10 pp 2081-2084.
4. CIE-publication W-1-1-1971.
5. Pirenne, M.H., Vision and the Eye (1968), (Chapmann & Hall, London).
6. O.C. Jones, M.A., D.Phil. I. Inst. (1978), Proposed Changes to the S1-system of photometric units, Lighting Research and Technology, Vol.10. No. 1 pp. 37-40.
7. Janssen, J.W., "S1-meter", Report Physical Laboratory TNO, The Hague.
8. Meijgaard, A.H.C. van, Shortwave-Infrared-Engineering, Part I - Part VI, IUB, The Hague.
9. British Standard, 4727; Part 4.
10. Meijgaard, A.H.C. van, 'Radiometry, Photometry and Receptometry, paper presented during the seminar 'Light measurement in industry', Guildford, Sept. 1978.

APPLICATION PARAMETERS OF A LIGHT SOURCE; LUMINOUS EFFICACY VS COLOUR RENDERING

J.J. Opstelten
Philips Lighting Division
Eindhoven
The Netherlands

The three most important characteristics of a light source intended for general lighting purposes are :

a) the luminous efficacy η defined by the quotient of the luminous flux ϕ_v emitted and the power P consumed by the light source,
b) the colour appearance specified by the colour temperature T_c or more precisely by the chromaticity coordinates x and y in the CIE 1931 standard colorimetric system and
c) the colour rendering properties specified by the general colour rendering index R_a.

The theoretical maximum of the luminous efficacy, η = 680 lm/W, will be achieved by a theoretical light source in which the power consumed is totally converted in monochromatic radiation at a wavelength of 555 nm. But this light source will not be acceptable for general lighting purposes because of its bad colour appearance and colour rendering properties.
Restricting the colour appearance of light sources to that of natural daylight and Planckian radiators the theoretical maximum of η lies between 380 and 520 lm/W, the exact value depending on the colour temperature T_c. These maximum values are achieved by 2-line light sources radiating at 445 nm and at a "yellow" wavelength complimentary to this "blue" wavlength. However the application of these 2-line light sources will be restricted due to their bad colour rendering properties. Good colour rendering properties combined with high luminous efficacies can be achieved with 3-line light sources radiating at three wavelengths, one in the blue (± 460 nm), one in the green (± 540 nm) and one in the red (± 610 nm) region of the visible spectrum.

A different approach of optimizing light sources which combine good colour rendering properties and high luminous efficacies starts with the spectral power distribution of light sources with maximal colour rendering properties. Examples of this approach will be given.

The foregoing theoretical considerations led to the realization of fluorescent lamps which are a great step forward in the compromise between high luminous efficacy and good colour rendering properties.

Literature

M. Koedam and J.J. Opstelten, Measurement and computer-aided optimization of spectral power distributions, Ltg. Res. & Techn. 3, 205-210, 1971.

W.A. Thornton, Luminosity and color-rendering capability of white light, J. Opt. Soc. Am. 61, 1155-63, 1971.

K. Mahr, Colour rendition and luminous efficacy. Particularly of combinations of four or five narrow band emissions, Extended Abstracts Spring Meeting Electrochem. Soc. 72-1 (Houston 1972), abstract No. 76.

J.J. Opstelten, D. Radielovič and J.M.P.J. Verstegen, Optimum spectra for light sources, Philips tech. Rev. 35, 361-370, 1975.

J.L. Ouweltjes, The general colour rendering index, its meaning and its use, Colour 73, York, B 213, p. 362-364.

Prime-Color Light Sources

W. A. Thornton
Westinghouse Electric Corporation
Bloomfield, New Jersey, USA

Light sources are generally used to ease human color vision. They should therefore be designed for human color vision. Human color vision is three-dimensional, implying three independent samplings by the visual system of each incoming light. These samplings must be represented by three different spectral system responses, and these system responses must in turn be characterizable by three wavelengths in the visible region of the spectrum (the three "prime-colors").

What are the prime-color wavelengths, which can be thought of as marking the sampling points of the human visual system? Sperling and Harwerth[1], and King-Smith and Carden[2], place them by direct measurement near 440 nm, 530 nm, and 610 nm. Indirect determination through psychophysical measurements places them near 445 nm, 535 nm, 615 nm.[3-6] The color-matching functions of the 1931 CIE Standard Observer, which unfortunately deviates importantly from human vision[7], place them near 450 nm, 540 nm, 600 nm[8].

Energy conservation demands that light sources be designed for maximum visual efficiency. If the prime-color wavelengths mark the maxima (or averages) of the three visual system responses, then it would follow that white light composed of the three prime-colors should yield maximum visual efficiency.

So far, only the fluorescent lamp can concentrate its output power in the three prime-color wavelengths. When that is achieved, three remarkably strong visual effects[9] result: (1) Perceived brightness per footcandle is observed to be substantially greater than that of "standard" or "deluxe" fluorescent lamps. (2) Clarity of the illuminated scene, in the sense of sharpness of detail, is substantially greater than that under standard or deluxe fluorescent lamps. (3) Attractiveness of coloration in the illuminated scene exceeds that of deluxe fluorescent lamps. All of these visual effects depend on the spectral power distribution of the illumination from the light source. They appear at present to be maximized when the lighting power is concentrated near the prime-color wavelengths. They appear to decrease sharply if yellow or blue-green wavelengths are allowed to form an appreciable part of the power in the illuminant.

Fig. 1 suggests a schematic of the overall spectral response[10] of the normal human visual system, inferred from all of the data of references 1-6. The overall system response is composed of three independent responses roughly centered on the prime-color wavelengths. As King-Smith and Carden[2] show, the three peaked responses shrink, under uncommon visual conditions, to the familiar luminous efficiency curve which defines the lumen and footcandle. The luminous efficiency response - undeniably active under certain visual conditions - should nevertheless be recognized to be a <u>sub</u>response, seldom if ever representative of normal color vision.

Fig. 2 shows the continuous spectrum (A) of average daylight and suggests the ideal matching prime-color illuminant (B) void of power in the violet, blue-green, yellow, and deep-red regions of the spectrum. Such an illuminant is designed for human color vision at minimum expenditure of energy. The accompanying clarity and coloration of the illuminated scene seem to be unrealizable by any other means; they are specifically absent from scenes illuminated by continuous spectra like daylight and incandescence. On the other hand, experimental fluorescent lamps have been used to show that the visual effects strengthen as the illuminant spectral power distribution changes progressively from a continuum toward the prime-color configuration of Fig. 2B.

Therefore, the visual performance of a prime-color illuminant may be progressively approached by any light source, by progressively concentrating its output near the prime-color wavelengths, to the extent that that becomes possible. Luminescent materials may be chosen and tailored for prime-color emission, for use with ultraviolet-emitting discharges. Arc constituents may be chosen and tailored for prime-color emission, in cases where most of the visible output is to be from the arc itself. In time, even the incandescent lamp may become alterable, and take on some of the prime-color character; as familiar and comfortable as incandescence lighting is, the achievement of even a little prime-color structure in its spectral power distribution enhances coloration and brightness.

Fig. 1

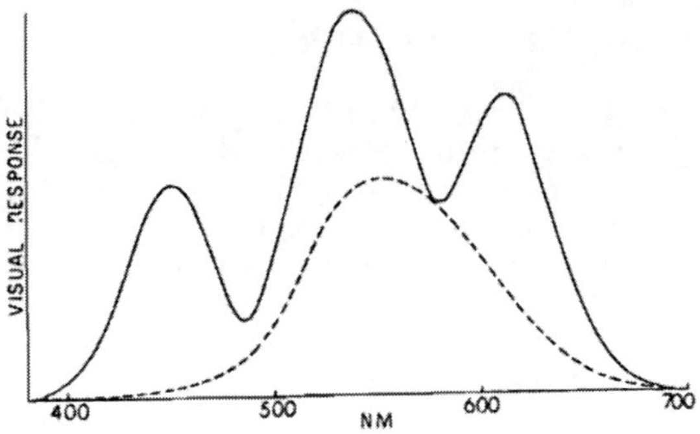

Fig. 2

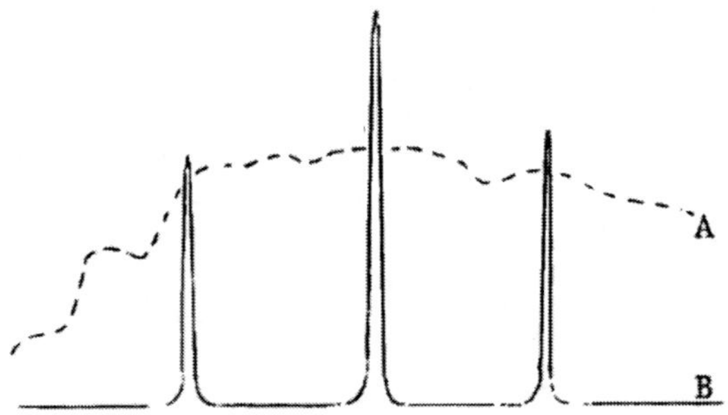

1. H. G. Sperling and R. S. Harwerth, Science Vol. 172, 9 April 1971, p. 180.
2. P. E. King-Smith and D. Carden, J. Opt. Soc. Amer. Vol. 66, July 1976, p. 709.
3. D. L. MacAdam, J. Opt. Soc. Amer. Vol. 28, April 1938, p. 103.
4. D. Jameson and L. M. Hurvich, J. Opt. Soc. Amer. Vol. 45, July 1955, p. 546.
5. W. A. Thornton, J. Opt. Soc. Amer. Vol.61, September 1971, p. 1155.
6. W. A. Thornton, J. Opt. Soc. Amer. Vol.62, March 1972, p. 457.
7. D. B. Judd, Ill. Eng. Vol. 53, February 1958, p. 61.
8. W. A. Thornton, J. Ill. Eng. Soc. (in press).
9. W. A. Thornton, Lighting Design and Application Vol. 5, November 1975, p. 35; Vol. 6, August 1976, p. 46.
10. W. A. Thornton, Colour 77 (Hilger 1978), p. 284.

MAPPING OF GaP LEDs PROPERTIES ON POLYCRYSTALLINE SSD SUBSTRATES

D. Nohavica, Institute of Radioengineering and Electronics, Czechoslovak Academy of Sciences, Lumumbova 1, 182 51 Praha 8, Czechoslovakia.

Substrates from bulk ingots of GaP prepared by SSD technique have relatively good structural properties for green and red GaP LEDs. Different orientations of grains may influence the impurity incorporation during LPE growth of p-n junctions. The dependence of the relative efficiency of LEDs prepared from different grains on one substrate is investigated by mapping technique. The dispersion of external quantum efficiency data (η_{ext}) is presented and compared with LEC GaP substrates with (111)B and (100) orientations. Obtained dependence of η_{ext} on grain orientations is discussed on the base of surface band bending or "available site" models of impurity incorporation.

Color Electroluminescence from ZnS Thin Films Multi-Doped with Rare-Earth Ions

S. Kurita, S. Takahashi, T. Sakaguchi

(Faculity of Engineering, Keio University, Yokohama, Japan)

It has been reported by D. Kahng in 1968[1] that efficient electro-luminescence (EL) is observed from thin films of ZnS doped with rare-earth ions. D. C. Krupka has measured the ratio of the emission intensities from the $^5D_3$ and $^5D_4$ levels of Tb^{3+} in the ZnS:Tb cells as a function of applied voltage and demonstrated that dominant mechanism for this EL devices is direct-impact excitation of rare-earth ions by not elecron[2].

The emission lines in this devices are obtained with corresponding to the orbital transitions of rare-earth ions and appeared individual color with weak coupling to the host ZnS lattice. Therefore, by means of color blending of different kinds of rare-earth ions, it may be available to yield the desired color emission from ZnS cell when the concentration ratio of rare-earth ion components is adequately selected.

The EL cells studied were prepared by simultaneous evaporation of ZnS and rare-earth fluorides from the independent sources. As shown in Fig. 1, this structure is composed of a ZnS layer doped with rare-earth ions and SiO layer sandwiched between Al contacts and basically similar to the one first reported by D. Kahng. All measurements were carried out in the pulsed operation at 270 Hz and pulse-width of 200 usec.

Fig. 2 shows three relationships of the concentration of rare-earth ions and relative brightness at fixed applied voltage. Optimum concentration are attained in the neighborhood of 2.0 at. % of Tb, 3.0 at. % of Sm and 2.0 at. % of Tm, respectively. Typical spectrum from the blend cell are shown in Fig. 3. This spectrum can be obtained when individual spectrum is summed with an adequate mixture ratio. The emission intensities at shorter wavelength side of the spectrum increase with increasing Tb concentration from 1.0 to 1.5 at. % as indicated by dashed curve in Fig. 3.

Fig. 4 shows relative emission intensities at each peak wavelength of two rare-earth ions from ZnS:Sm, Tm cell as a function of applied voltage. The emission intensities at 4800 A corresponding to $^1G_4 \rightarrow {}^3H_6$ transition in Tm^{3+} ions increase with higher rates compared with the emission at 6500 A for $^4G_{5/2} \rightarrow ^6H_{9/2}$ transition in Sm^{3+} ions when applied voltage arise. In addition, it found that the emission color of EL cell change when the applied voltage is changed.

The color coordinates of the emission from the both individual and blend cells are shown in the CIE chromaticity diagram in Fig. 5. This show also the resulthing color shift for four different applied voltages. This shift is attributed to the shift of the hot electron distribution in the ZnS films to higher energies with increasing voltage. Although color is independent of voltage variations, optimizing of blending ratio may be available to yield the desired color emission by cascading even if white color emission.

References

1) D. Kahng, Appl. Phys. Letters 13 (1968) 210.
2) D. C. Krupka, J. Appl. Phys. 43 (1972) 476.

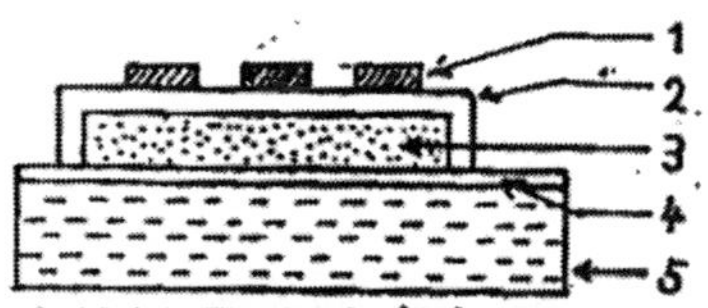

1. Metal Electrode (Al)
2. Insulating Layer (SiO)
3. Active Layer ($ZnS:LnF_3$)
4. Transparent Electrode (SnO_2)
5. Glass Substrate

FIG-1

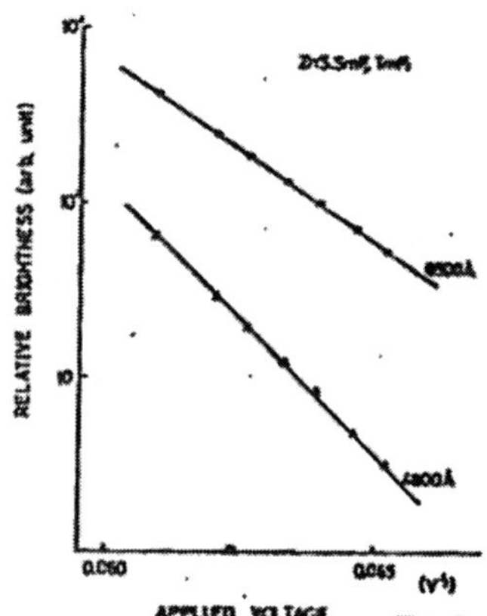

FIG-4

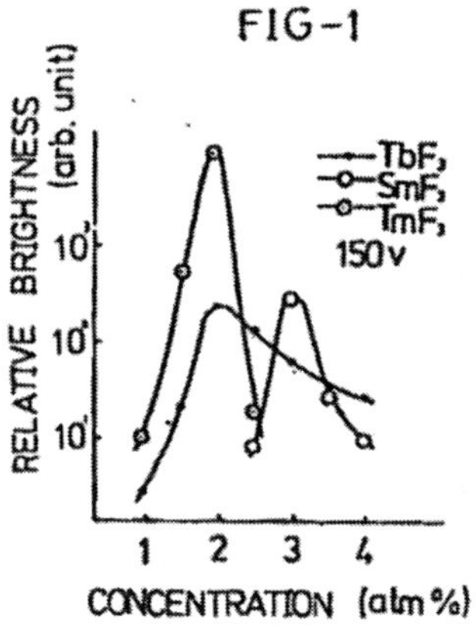

FIG-2

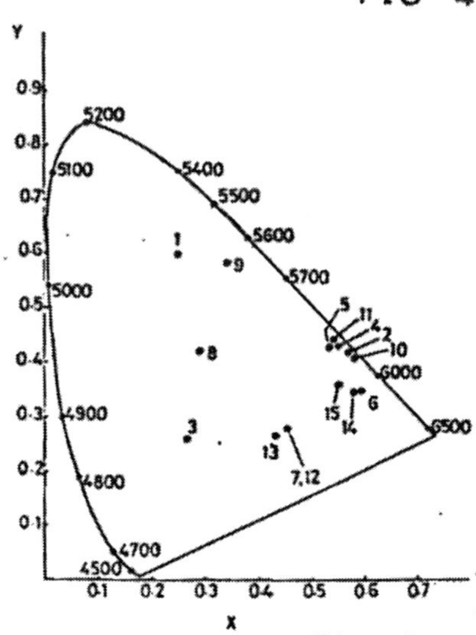

FIG-5

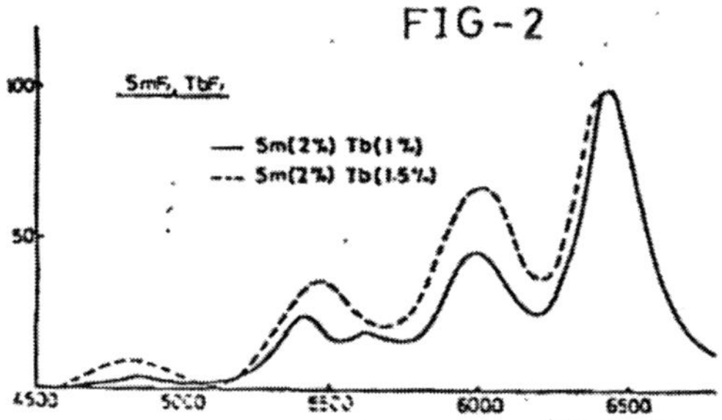

FIG-3

Incandescent Lamps.

L. Bartha
Res. Inst. für Technical Physics
Budapest, Hungary

At the centenary of the incandescent lamp discovered by Edison in 1879 it can be stated, that it has been one of the most successful industrial products of our modern age. The basic steps of the improvements carried out mainly in the last 60 years, which led to the present form of lamps are the following:

tungsten filament instead of carbon one
neutral gas atmosphere instead of vacuum
coiled filament instead of straight wire
halogen containing atmosphere instead of neutral one

As a result of these improvements both the life and the efficacy of the lamps increased considerably.

The light industry has always put great effort in the research therefore many processes occuring in the lamp are quite well understood. Some results of the recent years are:

The discovery of the potassium bubbles in the NS-tungsten wire gave a new explanation of the doping mechanism and of the high temperature mechanical behaviour of the tungsten wire.

The study of convection and transport phenomena gave a key to better lamp design.

The investigation of high temperature chemical reactions led to the development of the halogen lamps.

The main goal of the research at the present is the full understanding of life controlling processes and failure mechanisms.

It can be stated, that lamps burn out practically only at hotspots which develop for different reasons and reach the melting point of the tungsten. Hotspots can develop because of several reasons, like

a., original wire defects
  aa., diameter defects
  ab., resistivity defects
  ac., emissivity defects
b., original coil defects
c., defects caused by slow processes occuring during the life
  ca., bubble migration and coalescence
  cb., facet formation

cc., evaporation

cd., chemical processes

ce., secondary coil defects

There is no full agreement in the literature about the explanation, or even the acceptance of the above factors which probably means that under different conditions the importance of them might differ.

Beside of the efforts for better understanding and controlling the processes one can find new directions in lamp research and development some of which are very promising.

The work on the fluorine cycle may result a really selfhealing lamp with an increased life and efficacy.

The reinvestigation of the carbon filament proved that it might have some advantages if combined with a suitable chemical transport cycle.

The paper tryes to give a review on the results of the above topics based on the literature of the last few years.

The research activity, the wide field of application and the figures of lamp production give the impression that the second century of the incandescent lamp will be as bright as was the first one.

MORPHOLOGICAL EFFECTS OF TUNGSTEN FILAMENTS IN HALOGEN LAMPS: THEIR INFLUENCE ON THE TEMPERATURE DISTRIBUTION AND EVIDENCE FOR HOT SPOT GROWTH BY DIFFERENT FACETTING

S. Garbe
(Philips GmbH Forschungslaboratorium Aachen, Aachen)

The recrystallization of the tungsten filament and the tendency to form equilibrium surfaces with lower surface energy - a process which is known as facetting - changes the original temperature distribution of tungsten coils in certain halogen lamps. At constant voltage burning a temperature decrease of 100 to 150 K is observed at initial temperatures around 3000 K due to the increase of the emissivity of the facetted surface. Nevertheless, a constant luminous sensitivity is obtained throughout most part of the life.

The growth of low index facets (e.g. (110) with (211) edges) shows a $t^{1/3}$-time dependence and its temperature dependence is determined by the respective surface diffusion activation energies (80 - 120 K Joule).

Grain boundaries cause preferential evaporation with the formation of grooves. Dissimilar facet fields of neighbouring turns of a coil form local differences of the effective emissivity. The emissivity profile of a coil can be measured by simultaneous brightness and two colour pyrometry. Temperature differences as large as 50 K develop due to effective emissivity gradients. Such temperature gradients between adjacent turns of a coil increase the axial transport of tungsten and lead to hot spot formation. The reduction of life may be as large as a factor of 2.

Cross sectioning of coils at the end of life by polishing and planimetry give an indication of the degree of facetting and notching. Together with grain boundary effects and the formation of large potassium bubbles facetting and notching may lead to fracture of the wire and premature lamp life.

29.

CALCULATION ON THE DEVELOPMENT OF HOT SPOTS IN STRAIGHT WIRE TUNGSTEN FILAMENTS CAUSED BY AXIAL MASS TRANSPORT

J.J.F. Geijtenbeek
Philips Lighting Division
Eindhoven
The Netherlands

If a straight tungsten wire is heated by an electrical current up to temperatures above 3000 K in an inert gas atmosphere very often burn-out takes place near the cold ends of the wire. This is assumed to be caused by hot spot development due to axial mass transport.

In order to verify this assumption calculations have been made to determine the influence of axial mass transport on hot spot development.

Starting from the energy balance for the wire at time t=0, successive changes in the geometry of the wire are calculated in time intervals $\Delta t << \tau$, where τ is the life of the filament. The end of life is defined as that time when a local temperature of the wire exceeds the melting point.

In each step the temperature distribution and axial and radial tungsten transport are calculated using the Langmuir film model assuming steady state conditions.

The calculations indicate that for a wire without defects burn-out takes place near the ends of the wire. For a wire with an initial defect, burn-out at the position of this defect only takes place if the defect exceeds a critical magnitude, otherwise burn-out occurs near the cold ends.

MIGRATION OF CARBON IN LINEAR TUNGSTEN HALOGEN LAMPS

I. CONNOR BSc MSc

Thorn Lighting Limited, Research and Engineering Laboratories, Leicester, England

Although not fully understood, the role of carbon in tungsten halogen lamps has only been considered as one of oxygen gettering[1][2]. In the experiments described here the presence of carbon in an iodine lamp has had a detrimental effect on lamp performance.

During an investigation into the use of solid dosing with iodoform (CHI_3) for 1000W linear tungsten halogen lamps, unexpected blackening of the bulb wall was encountered. This appeared early in life between the first two filament supports in the end of the lamp dosed with halogen. The blackening did not affect all lamps. Only a small percentage of the lamps run in enclosed fittings exhibited this specific type of blackening.

By conducting a series of simple experiments, where the CHI_3 was moved within the lamp prior to initial ageing, it was established that the blackening occurred near to wherever the CHI_3 had been concentrated. Considering the elements present, carbon was suspected of being the component inducing blackening.

A programme of experiments was begun to trace the movement of carbon within the lamp, and also to try and gain information about the chemical reactions which take place. Radiochemical tracing was used for assessing carbon movement, and iodoform containing carbon-14 was made at Loughborough University for use in experimental lamps.

Lamps were made with a deliberate CHI_3 overdose to ensure blackening. These were split into two groups, one for radiochemical study and the second for analysis by Auger electron spectroscopy.

The distribution of carbon found is plotted in Figures 1 and 2. These show the carbon-14 per unit length of lamp, for lamps aged only and lamps run for 65 hours. The activity measured after 65 hours was considerably greater than that found after ageing, see Table 1.

TABLE 1 - TOTAL ACTIVITY FOUND IN TEST LAMPS

	After Ageing	65 Hours
Filament activity	0.051 µCi	0.083 µCi
Bulb activity	0.005 µCi	0.102 µCi
Total activity	0.056 µCi	0.195 µCi
Theoretical total activity	0.192 µCi	

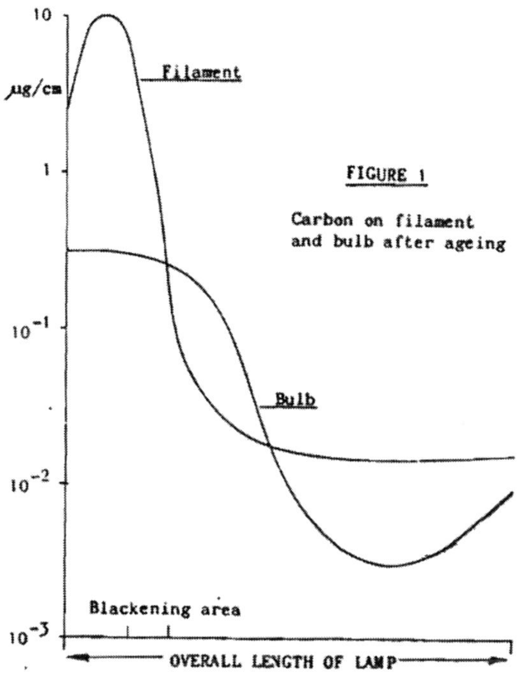

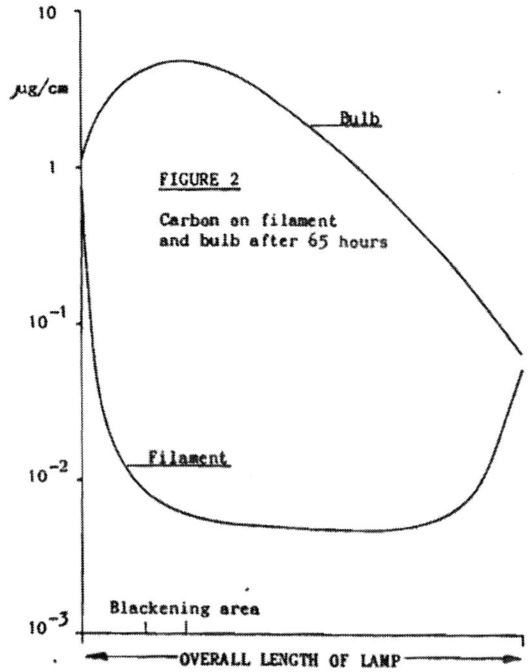

Auger electron spectroscopy of the filament section nearest to the site of blackening confirmed that the filament was the source of the increased activity. Carbon levels in filaments after ageing were significantly higher than those from a filament section taken from a lamp run without the CHI_3 dose. However within 65 hours the carbon found in filaments from dosed lamps fell below that from the undosed lamp. Auger electron spectroscopy of blackened bulb sections showed the stains to be a mixture of tungsten, carbon and tungsten carbide.

The actual mechanism for the blackening remains unclear, but three possible alternatives can be postulated:

(a) Tungsten carbide evaporated from the filament prevented the effective operation of the iodine cycle.

(b) Carbon evaporated from the filament competed favourably with iodine to form tungsten carbide, again reducing the effectiveness of the halogen cycle.

(c) Carbon from the filament formed carbon monoxide and effectively suppressed the iodine cycle.

Further experiments are being carried out to determine more details of the role of the carbon in this lamp situation.

REFERENCES

1. C. Collins and E.G. Zubler, U.S. Patent 3132278, 1964.

2. G.R. T'jampens and M.H.A. Van de Weijer, Philips Technical Review, 27, No. 7, 1966, pp 173.

THE ROLE OF METALLIC IMPURITIES IN THE LINEAR QUARTZ BROMINE LAMPS

L. N. Yannopoulos and A. L. Wolfe
Westinghouse Research & Development Center
Pittsburgh, Pennsylvania, U.S.A.

The achievement of steady light output and long life in the halogen-dosed tungsten incandescent lamps depends on the conditions that govern the chemistry of the tungsten-bromine regenerative cycle. For the desired long lamp life, especially in the case of the linear quartz bromine-dosed lamps, the cycle is required to function at low concentrations of the bromine compound added to the inert fill gas. It is well known that this can be realized only if a stringent control on the variables that influence the cycle is exercised. In our previous studies [1,2] and in a series of papers by G. Neumann [3], the role of many of these variables has been examined. The role of one of the variables, specifically that of the common bromine reactive metallic impurities, has not yet been reported. This paper deals with some experimental and theoretical aspects of this problem in the methylene bromide-dosed linear quartz bromine lamps.

On the basis of thermochemical considerations alone, it can be predicted that the tungsten filament stabilizing additives (Al, K, Si) and metallic residual impurities (Fe, Ni, Cr, Cu, etc.) can be released into the bulb of an operating halogen lamp. The extent of their release will depend only on their individual rates of diffusion from the bulk of tungsten to its surface. Once these metallic impurities are released into the bromine reaction space of the bulb, they can participate in the bromine cyclic reactions. In Figure 1 the temperature dependence of the vapor pressure data of the most stable bromides of these metallic impurities is compared with that of the bromides and oxybromides of tungsten. The figure clearly shows that, at the measured minimum bulb wall temperature ($\sim$300°C) of these long linear lamps (i.e., 500 W T3 Q/CL), the formation of condensed bromides of these metallic impurities is possible and can seriously impair the tungsten cycle if the amount of bromine in the fill gas is not in excess of the sum of the stoichiometric amounts in the impurity bromides.

These predictions were tested at an optimum cold argon fill gas pressure of 3 atm and at a bromine concentration range of 8.75-52.5 micrograms (μg) in 142 lamps of the type 500 W T3 Q/CL. Of this total, 54 contained 17.5 μg of bromine. The details of lamp filling and testing procedures have already been described [2]. In all tested lamps, the accumulation of bromine-containing deposits at the corners of the fused quartz bulbs was observed. In several lamps the cycle was not sustained and the lamps blackened early in life. The lamps that continuously burned clean reached an average useful life close to 4000 hours. In the case of the blackened lamps, the minimum bulb wall temperature, $t_{b(min.)}$, was locally increased from $\sim$300°C to $\gtrsim$500°C using small heat reflectors around the corners of the lamps. Continuous burning with $t_{b(min.)}$ $\gtrsim$500°C resulted in the elimination of blackening and in the maintenance of steady light output to the end of the life of the filament. This average life was also close to 4000 hours.

In typical lamps, the deposits were analyzed and found to contain bromine, iron, nickel, chromium, and copper. Spark source mass spectrography and emission spectroscopy were used for the identification and determination of the constituents in the deposits. Emission spectroscopy was also used in monitoring the release of metallic impurities during lamp burning. The spectrograms were obtained as a function of lamp burning time and indicate interesting changes in the concentration of the excited impurity atoms in the vicinity of the filament. The absence of K emission lines in the spectrogram is noteworthy.

The observed behavior of the tungsten-bromine cycle is explained from the chemical analysis data of the constituents of the bromine-containing deposits and their observed vaporization behavior at ∿300°C and ≳500°C. The explanation is consistent with the vapor pressure data of the bromides in Figure 1.

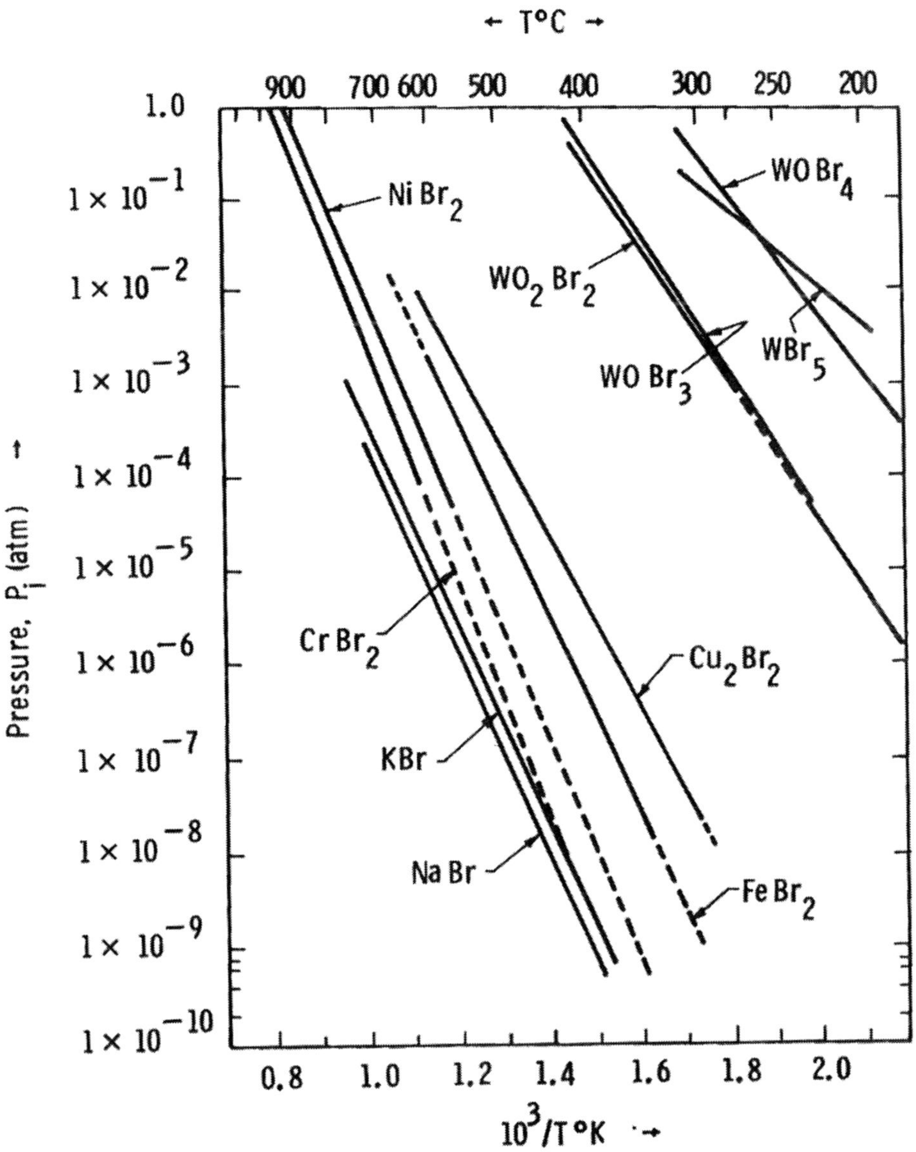

Fig. 1 — Log P_i versus $\frac{1}{T\,(°K)}$ of stable bromides of metallic impurities and Tungsten in quartz bromine lamps

References

1. L. N. Yannopoulos and A. R. Pebler, "Direction of Chemical Transport of Tungsten in Tungsten-Bromine Lamps", J. Appl. Phys., 43, No. 5, 2472 (1972).

2. L. N. Yannopoulos, "Experimental and Theoretical Considerations of the Life Limiting Factors in Quartz-Bromine Lamps", J. Illum. Eng. Soc., 5, No. 2, 74 (1976).

3. G. M. Neumann, "High Temperature Thermodynamics of Chemical Transport Reactions in Tungsten-Bromine System of Halogen-Incandescent Lamps and its Influence by Hydrogen, Oxygen, and Carbon", Thermochemica Acta, 8, 369-379 (1974). (See previous references in this paper.)

ANALYSIS OF TUNGSTEN COMPOUNDS IN THE WALL REGION OF BROMINE HALOGEN LAMPS

P. Eckerlin and S. Garbe
(Philips GmbH Forschungslaboratorium Aachen, Aachen)

In bromine halogen lamps with relatively high doses of HBr or CH_2Br_2 and some residual water vapour sometimes precipitates of tungsten oxibromides are formed at comparatively low temperatures of the silica bulb. WBr_4 or $WOBr_4$ are often assumed to form in the wall region.

The question, whether local thermodynamic equilibrium is obtained at the bulb wall, can be answered, if the dominant species and their concentration can be measured.

In order to detect the respective compounds, H 3 single coil lamps with an attached small capillary of 300 µm diameter and 2 cm length were well degassed, processed, and filled with HBr and argon.

During lamp life the capillary had a temperature of 600 K at the sealing to the bulb and a temperature of 500 K at the end of the capillary. Under the influence of this temperature gradient the capillary acts like an effusion cell and the stable compounds in the gaseous phase are forced to condense at the respective places of the capillary, where their density is larger than the respective vapour pressure of the compounds.

At the end of life the capillary was sealed from the lamps and X-ray diagrams were taken in a Debye camera. WO_2Br_2 , $WOBr_2$ and W_6Br_{18} were obtained in the intensity ratios 40:1:0.1 in lamps which had normal life with no precipitates formed in the bulb itself. The addition of 2.10^{-2} bar CO increased the formation of the precipitates. The amount of WO_2Br_2 was 10 µg and corresponded to a pressure of $2 - 5 \times 10^{-5}$ bar in the bulb wall region. Lamps filled with HCl (5×10^{-2} bar, 1 bar argon) showed precipitates of WO_2Cl_2 , but no indication of $WOCl_4$ or $WOCl_3$ was obtained.

The experiments are in agreement with thermodynamic equilibrium calculations.

EVALUATION OF THERMODYNAMIC DATA OF TUNGSTEN BROMIDES AND OXYBROMIDES AT HIGH TEMPERATURES

G. Dittmer and U. Niemann
(Philips GmbH Forschungslaboratorium Aachen, Aachen)

In order to get insight into the relevant processes occurring in halogen-filled incandescent lamps chemical equilibrium calculations have been carried out. Calculations are based on the assumption that the lamp volume is divided into concentric equilibrium compartments between the filament and the bulb wall. In combining the law of mass action, the law of mass conservation within the bulb volume and the balance of diffusion-controlled mass fluxes between each of the equilibrium compartments the total amount of tungsten dissolved in the gaseous phase is calculated.

Tungsten solubilities were obtained experimentally on systems containing solid tungsten in contact with an inert gas and small amounts of $Br_2/O_2/H_2$-mixtures. In large glass bulbs of 300 ml volume tungsten coils were heated until the current through the coil reached a constant value. The respective temperatures were adjusted pyrometrically. From the mass loss of the coil the total tungsten pressure was evaluated according to

$$SPW = \frac{\Delta m \cdot R \cdot T}{M_W \cdot V}$$

SPW total tungsten pressure
M_W atomic weight of tungsten
V volume of the bulb
R gas constant
T temperature
Δm mass loss of the coil

During experiments the glass bulbs were kept at 700 K by external heating to prevent equilibrium compounds from condensation. This failed, however, in experiments with $H_2:Br_2$ ratio of $\gtrapprox 1$ where the deposition of solid tungsten oxides was observed.

In the binary W/Br_2 system kinetic effects did not allow an accurate determination of the amount of tungsten dissolved in the gaseous phase at temperatures below 1500 K.

In the way outlined above estimated thermodynamic data could be revised in adjusting calculations to experimental results. Thermodynamic data finally used in the equilibrium calculations will be given.

According to heterogeneous equilibrium calculations of the $W/Br_2/O_2/H_2$ system the dominating tungsten species change from WBr_4 , WO_2Br_2 WO_2Br to WO_2 with increasing temperatures of the filament.

TAIL-EROSION IN HALOGEN INCANDESCENT LAMPS : KINETICS OF THE W - Br - O - H - SYSTEM

G. Rouweler and B.J. de Maagt
Philips Lighting Division
Eindhoven
The Netherlands

The rate of reaction of a polycrystalline tungsten wire was determined at atmospheric pressure in a flow of argon gas, containing bromine (1), bromine and oxygen (2), or bromine, oxygen and hydrogen (3). The rate was measured as a function of gas composition and temperature by monitoring the electrical resistance of the wire (lit. 1). By applying isothermical conditions and adapting the gas-flow velocity, mass-transport effects in the gas phase were minimized.
The kinetics of the first system can be expressed by the empiric relation (j in mol cm^{-2} s^{-1}) :

$$j_1 = 4.7x10^{-13} \times p_{Br} + 4.0x10^{-15} \times p_{Br}^2 \quad (10<p_{Br}<10^3 Pa;\ 975<T<1282K)$$

It is supposed that the rate-limiting step in this reaction is the addition of a bromine atom from the gas phase to a tungsten-bromine surface complex (Eley-Rideal mechanism). The second system shows a more complicated reaction behaviour :

$$j_2 = k_2 \times p_{O_2}^f \times p_{Br_2}^g$$

T(K)	p_{O_2} (Pa)	p_{Br_2} (Pa)	$10^{10}.k_2$	f	g
1067	1 - 30	0.3 - 3.10^3	9.8	1.55	-0.92
1166	0.25 - 8	2.10^{-2} - 3.10^3	12	1.20	-0.61
1282	6.10^{-2} - 8	10^{-2} - 3.10^3	23	1.05	-0.40

Here the reaction rate, which decreases with increasing bromine content of the gas phase, is limited by the competitive adsorption of oxygen on the surface, which is nearly completely covered with bromine (Langmuir-Hinselwood mechanism).

In the third system with HBr and H_2O as the main gas-components two reaction regions are observed depending on the gas composition. When bromine is the excess component then the reaction follows the mechanism of the second system. When hydrogen is the excess component a complicated reaction pattern is observed for which, within the range of our experiments, no theoretical explanation can be given.

Lit. 1 : G. Rouweler, B.J. de Maagt
Philips J. Res. 33, 1 (1978)

Gettering of Incandescent Lamps by a Chemical Conversion Technique

J. H. Singleton
Westinghouse R&D Center
Pittsburgh, Pennsylvania 15235 USA

The economic production of incandescent lamps necessitates the use of very high speed production equipment, which reduces the time available to achieve a satisfactory vacuum or gas-fill environment in the lamp. In a typical commercial lamp the major source of filament erosion is water vapor, with a smaller contribution from carbon dioxide. These impurities are present in the glass envelope and their amounts can be very markedly increased by an internal light-diffusing coating, such as a silica powder. It is possible to achieve satisfactory residual levels of impurities by a standard evacuation/flush technique, at elevated temperatures, but it is common practice to also incorporate a chemical getter, such as those based on zirconium, into the sealed lamp. An alternative technique to the standard processing, which can give good lamp performance even with inadequate conventional processing, is to employ a chemical conversion process by which the oxidizing impurities are converted to carbon monoxide and hydrogen. It has been clearly demonstrated that, at tungsten temperatures above ∿2200°K, the reaction of oxygen with tungsten which contains carbon impurity preferentially leads to the desorption of carbon monoxide rather than tungsten oxides.[1] Even when the carbon has been removed from the filament the desorbing species shifts to atomic oxygen[2] and this has a sufficient lifetime to react with carbon-containing species elsewhere in the lamp envelope. These reactions also occur with water vapor and with carbon dioxide. The simplest demonstration of the technique is to dope the fill gas with ∿0.1% methane, or similar gas. In a test of this procedure one group of lamps was prepared using methane doping and compared to one using standard fill gas. Lower than normal processing temperatures were used, and the lamps were burned in a fixture such that the wall temperature was 400°C, so as to accelerate the release of water. The conventional fill gas lamps gave a mean life of 23 $\pm$ 8 h (range 16 to 38 h), and the methane doped lamps gave a mean life at 99 $\pm$ 151 h (range 18 to 719 h).

Two problems can arise with the addition of methane. The flashing of the filament to achieve a stable grain structure causes decomposition of ∿50% of the methane, incorporating some carbon into the filament which can give embrittlement. Secondly, above a critical concentration (∿1% CH_4 in these tests) the flashing initiates a very rapid reaction with the methane, leading to filament distortion or destruction. While proper adjustment of the process can avoid these problems, a more viable application of the chemical conversion process is to incorporate a stable compound having a high carbon content, into the wall coating. Interestingly, one technique of producing a silica coating, by burning ethyl orthosilicate[3] often yields such trace amounts of organic materials, because of the difficulty of achieving complete combustion of the orthosilicate. A number of organic materials have been successfully tested for this application, the major requirement being for a low vapor pressure at the maximum tolerable temperature of the operating lamp in an enclosed light fixture. For example a group of lamps in which 5×10^{-4} g of stearic acid was incorporated on the wall coating were compared to undoped lamps, again after processing at lower-than-normal temperatures. When the lamps were burned at 400°C wall temperatures the controls gave a mean life of 28 ± 23 h (range 8-34 h) and the doped lamps 904 ± 400 h (range 132-1321 h). Using this technique there is little danger of filament embrittlement since the reaction of oxygen atoms occurs at the walls and the vapor pressure of the organic dopants is negligible under all normal lamp operating conditions.

1. J. H. Singleton, J. Chem. Phy. 45, 2819 (1966).
2. P. O. Schissel and O. C. Trulson, J. Chem. Phys. 43, 737 (1965).
3. M. Pipkin, U. S. Patent 2,545,896, March 20, 1951.

LAMP MATERIALS

J,R. COATON

Thorn Lighting Limited, Research & Engineering Laboratories,

Leicester, England.

The history of artificial lighting clearly shows that the development of devices has been strongly dependent upon the availability of suitable materials. This has applied from the gas mantle through to the most complex multi-metal halide gas discharge lamp. The required properties of lamp making materials are many and varied including melting and working temperatures, strength, ductility, weldability, corrosion and oxidation resistance, expansion, conduction resistivity, finish, vapour pressure, etc and even the ordinary general lighting service incandescent lamp has some twelve components, all carefully chosen for their physical and chemical properties.

The most widely used envelope material is soda-lime-silicate glass, and since it is adequate and inexpensive it is unlikely to be replaced by an alternative. However, for economic and ecological reasons it may be necessary to adopt lower lead (or lead free) glass for stems and miniature lamp bulbs.

Vitreous fused silica (quartz) is an important material for making discharge lamp arc tubes and tungsten halogen lamp envelopes. This was principally made by fusing rock crystal, but the traditional source of raw material was virtually priced out of the market during the early 70's, and many manufacturers developed an alternative process using the more readily obtainable pegmatitie material, fused in a helium/hydrogen atmosphere. Such material, after suitable baking, may have a hydroxyl content of less than 1ppm. However, its physical properties differ in some respects from the long established fused quartz, and further research is needed to assess the properties of this so called 'sand based' material.

The high pressure sodium vapour discharge lamp is a good example of the important role of material for lamps. In this case the effect of high pressure filling was predicted and experimentally observed long before a sufficiently refractory envelope material was available. The discovery in 1957 that aluminium oxide could be sintered to give a dense, transparent material, having a high resistance to chemical attack and with a melting point of 2050°C, opened the way for the development of this important light source. Much effort has also been directed into developing suitable hermetic seals, adding much to the reliability of the device. Speculative work again indicates the growing need for a more refractory material than alumina; yttria and spinel have been proposed and both merit further study.

Research and development on tungsten has been simulated by requirements in two principal areas:-

a) Coiling machine development, where increased winding speeds has dictated the mechanical properties of the 'as drawn' wire.

b) Tungsten-halogen lamps, where the high filament operating temperature and extended life has set new standards for chemical purity and long term mechanical stability.

This latter aspect has lead to extensive investigations into the failure mechanism of filaments, and in particular the nucleation and growth of pores, which are essential for producing the desired structure, but can also contribute to premature failures in the later stages of their growth.

Much has been achieved in improving the uniformity and stability of tungsten wire during the last decade, which has benefitted filaments and discharge lamp cathodes, but there is still scope for improvement, and for example, the possibility of providing the essential micro-voids by implanting helium ions has been postulated, but has yet to be realised.

Metal halide gas discharge lamps and tungsten-halogen incandescent lamps are both important light sources capable of further development. However, the constituents in both are known to react with the vitreous silica arc tube/envelope (all-be-it very slowly) and this could prevent these important lamps from being developed to their full potential. Polycrystalline ceramic has been considered as a replacement to vitreous silica, but the niobium leadthrough and glassy sealing material react with the vapours present in the lamp during its operation. Various alternatives have been suggested, including the deposition of an inorganic glassy film from an organic solution, to form a thin protective layer; this has already proved effective in producing fluorine resistant coatings on vitreous silica components. Also, certain glass ceramics may have the required properties, but at present are difficult to fabricate.

In the field of incandescent lamps there is still the tantalizing prospect of raising the efficiency of this supremely simple device. Several alternatives have been suggested from time to time including materials other than tungsten, the application of double quantum phosphors to convert some infra-red radiation to light, infra-red reflecting light-transmitting bulb coatings to raise the overall lamp efficiency and the effective supression of filament evaporation by a genuine regenerative tungsten transport cycle. Progress is being made on at least two of these, and with a world potential in excess of 10^{10} lamps per year, a break through would certainly create a considerable impact on our energy and cost conscious society.

In conclusion, the development of light source, past and present, has been strongly linked to the availability of suitable materials to satisfy the many and varied requirements of electric lamps. The future is not likely to prove an exception and progress on discharge, incandescent and tungsten-halogen lamps is expected to proceed in conjunction with developments in materials.

ENERGY SAVINGS FOR AN INCANDESCENT LAMP FILAMENT IN A RADIATION REFLECTIVE SURROUND

Jack Brett, Raymond Fontana, Jim Parascandola, Wolfgang Thouret, Peter Walsh, Steven Spura and Luke Thorington.

The energy savings from the use of a reflective coating returning radiation to a compact high emissivity (ϵ = .55) coiled coil incandescent tungsten filament have been studied in a model system and with fabricated lamps. The highest conservation measured was 65% for a model light bulb system consisting of an optically polished, silvered, spherical, 76 mm diameter cavity with a perfectly centered filament supported on a micropositioner in argon.

Optical centering was accomplished by coincidence of the real and reflected filament image which was located in the same position as filament placement for maximum electrical resistance or maximum two-color pyrometer temperature at constant filament voltage.

Externally silvered 80 mm diameter spherical bulbs (Corning Glass Works G25, lime glass) conserved on the average 35% of the energy required to heat the filament to a given temperature in argon. However, the best performing silvered bulbs save up to 55% which is attributed principally to differences in the optical quality of the bulbs.

Lamp envelopes with an RF-sputtered heat-mirror coating of TiO_2-Ag-TiO_2[1] which transmits visible wavelengths while reflecting infrared radiation showed distinct improvement over envelopes with a simple TiO_2 coating[2] for energy conserving lamps.

Measurements of filament resistance as a function of wattage with such heat-mirror coated bulbs compared with uncoated bulbs have been made in the model lamp system. They indicate wattage reductions of up to 40% to achieve a given filament resistance. Concomitant reductions of at least 10% occur in horizontal candlepower due to absorption by the coating. Lamps have been constructed with the coating applied either externally or internally to standard G25 envelopes. Compared to uncoated lamps up to 25% less wattage was required for the same lumen output at equal filament temperatures. Measurements have been made of the differences between the temperature distribution of coiled coil filaments in reflective enclosures and that in standard clear bulbs. The influence of filament misalignment on the obtained wattage saving has been measured in silvered reflective spherical cavities.

(1) John C. C. Fan, F. Bachner, Applied Optics 15, 1976, p. 1012.

(2) Frank Studer, D. Cusano, Journal Opt. Soc. Am., 43, 1953, p. 522.

SAG AND TILT OF INCANDESCENT COILS AT COMPARATIVELY HIGH TEMPERATURES

H. Brons, W. van Hoorn and K.C. Thompson
Philips Lighting Division
Eindhoven
The Netherlands

The ever present gravitation makes the life of incandescent lamps highly dependent on condition and behaviour of the coiled filament.
Due to its own weight the coil is continuously subjected to mechanical forces which may give rise to two phenomena : "sag" and "tilt".
Sag results from creep under the influence of a mechanical load and is among other things, dependent on the grain size of the filament material. This phenomenon, which affects maintenance and lamp life, can be avoided by enlarging the grain size.
However, a different coil defect then arises. The coil can distort as a result of twisting of two adjacent wire parts with respect to one another, along a common grain boundary.
This causes a discontinuity in the coil pitch which normally results in short, sometimes quite short lifes.
The occurrence of this phenomenon "tilting" is dependent on Tungsten morphology. At sufficiently large grain size tilt is found, within 50% of nominal coil life, for all coils tested.
Calculations of sag, as a function of grain size have been performed in order to find the minimal grain size at which sag is confined to an acceptable level.
These calculations have been compared with observations of coils burning at comparatively high temperatures.

A FIELD-ION MICROSCOPE STUDY OF THE MICROSTRUCTURE OF TUNGSTEN FOLLOWING HELIUM ION IMPLANTATION

J.M. Walls and R.J.K. Nicholson

Department of Physics,
Loughborough University of Technology,
Loughborough, Leicestershire, LE11 3TU, U.K.

ABSTRACT

The field-ion microscope has been used to study the microstructural changes in tungsten following low-energy (1 to 3 keV) helium ion bombardment with doses in the range 10^{15} to 5×10^{17} ions cm^{-2}. These experiments have enabled the nature of the surface and sub-surface damage to be defined on the atomic scale. In addition, since the direction of bombardment has been along the axis of the field-ion emitter, subsequent field evaporation has allowed the precise depth distribution of the defects to be determined.

Vacancies and self-interstitial atoms have been observed in all the experiments. At low doses, viz 5×10^{15} ions cm^{-2}, the damage has been found to consist mainly of small bubbles and dislocation loops. At higher doses, viz 10^{17} ions cm^{-2}, substantial surface damage has been observed. The size, shape and distribution of the bubbles together with the density of dislocation loops has been determined as a function of ion energy and dose. The study has also shown that the depth of the gas bubbles is, in all cases, substantially greater than the depth predicted theoretically for the low energies employed.

Helium ion implantation at higher energies has considerable potential in the controlled introduction of gas bubbles in tungsten as a means of pinning the grain boundaries in the as-drawn structure thus producing good ductility at higher temperatures. In addition, implantation could also be used to promote elongated grain morphology during secondary re-crystallisation. Such micro-structural properties would promote high temperature strength and could lead to a prolongation of filament life.

PASSIVATION LAYER OF ALUMINIUM FLUORIDE PROTECTS FUSED QUARTZ LAMP ENVELOPES FROM ATTACK BY FLUORINE

J.R. FITZPATRICK AND V.W. GODDARD

THORN LIGHTING LTD, RESEARCH AND ENGINEERING LABORATORIES,
LEICESTER, ENGLAND.

A major problem in the development of a lamp based on a chemical vapour deposition process involving fluorine is the need to protect the fused quartz envelope from attack by fluorine. At elevated temperatures fused quartz will react with metal halides[1] in reactions of the type:

$$SiO_2 + \frac{4}{n} MX_n \rightarrow SiX_4 + M_{4/n}O_2 \qquad (1)$$

where MX_n represents the metal halide. Reactions of this type have been shown to take place in lamps containing fluorine at temperatures as low as 450K[2]. SiF_4 is formed coincident with blue tungsten oxide $W_{18}O_{49}$ appearing on the bulb wall. A black tungsten deposit forms on the bulb wall rendering the lamp useless after only a few minutes operation.

Solution - deposited glass coatings based on the system Al_2O_3-TiO_2-P_2O_5 have been developed[3] for use on the internal surfaces of lamp envelopes to protect the fused quartz from attack by fluorine. These coatings are thermally stable up to 900K, the maximum temperature reached by the envelope during operation. They have high transparency to visible radiation. Experimental lamps whose internal surfaces had been coated with this type of coating and which were then dosed with fluorine have been run for over 100 hours and remained clean.

The use of such coatings has enabled the development of lamps based on the fluorine transport cycle to be undertaken. From this work experimental analytical evidence has been obtained which leads to the conclusion that protection of the lamp envelope from attack by fluorine is due to the formation of a passivation layer of aluminium fluoride by a reaction between the aluminophosphate glass coating and fluorine contained in the lamp atmosphere.

Ion selective electrode techniques have been used to measure the amount of active fluorine in the lamp atmosphere after lamps had been run for times between 15 minutes and 100 hours (end of life). These measurements showed that there is a distinct loss of active fluorine through life, from an initial dose of around 50 μg to about 1 to 2 μg at end of life. This apparent loss of fluorine can be readily explained by a reaction occurring between the fluorine and the coating. The most likely reaction being:

$$2Al_2O_3 + 6F_2 \rightarrow 4AlF_3 + \text{"}3O_2\text{"} \qquad (2)$$

The oxygen reacts to form either tungsten oxides or oxyfluorides. Fluorine combined as part of the aluminium fluoride is not measured by the ion selective electrode technique as active fluorine.

Examination of the internal surfaces of lamps, after they had been run, by Auger electron and X-ray photoelectron spectroscopy (ESCA) showed clearly the presence of fluorine. ESCA revealed two forms of aluminium, one the oxide Al_2O_3, the other believed to be aluminium fluoride AlF_3.

The ESCA spectrum of pure aluminium fluoride has been measured and compared with the spectrum obtained from the coating surface to confirm the presence of aluminium fluoride on the surface. Electron diffraction experiments were carried out to identify the crystal structures existing on the glass surface.

Solid dosing techniques are used to maintain accurate control of the initial fluorine dose and a repeatability of ± 15% is possible(4). Gas dosing techniques result in a repeatability of dosing no better than ± 50%. The formation of a passivation layer requires that the initial dose must be considerably greater than the amount required to maintain a chemical transport cycle. This may result in a wider variation in the amount of active fluorine and ways of controlling active fluorine are under investigation.

REFERENCES

(1) Schafer, H., Chemical Transport Reactions (Academic Press) 1964.

(2) Haigh, I., Ph.D. Thesis, Leicester University (1973).

(3) Rothon, R. and Rees, J.M., Chemistry and Industry, 1 July 1978.

(4) Fitzpatrick, J.R. and Rees, J.M., Lighting Research and Technology, (to be published).

SINTERED YTTRIA FOR LAMP ENVELOPES

by

William H. Rhodes
GTE Laboratories, Inc.
40 Sylvan Road
Waltham, Massachusetts 02154
U.S.A.

A number of the physical properties of Y_2O_3 suggest that this material may prove an attractive lamp envelope. It should be possible to raise the operating wall temperature of the high pressure sodium lamp compared with the standard Al_2O_3 envelope because of the higher melting point (2464°C) and lower vapor pressure. The infrared transmission extends to longer wavelength, and suggests that the emissivity could be lower than that of Al_2O_3. Thermal shock resistance should be adequate for this application because of the low thermal expansion coefficient ($\alpha_{25-1000^oC} = 7.9 \times 10^{-6}\ ^oC^{-1}$) combined with moderate mechanical strength and low elastic modulus. Finally, high density polycrystalline samples should be free of internal light scattering from grain boundaries due to the cubic crystal structure.

Detracting from these virtues is a history that yttria-based ceramics, containing 2 to 15 mole % ThO_2, darken, presumably by loss of oxygen, in the environment of a high pressure sodium lamp at 1200° - 1400°C. The present study found instead that pure Y_2O_3 is more stable in this environment. Thus, the problem was reduced to finding sintering aids which permitted the attainment of translucent or transparent polycrystalline products, while the thermal stability of undoped Y_2O_3 was retained. Three such sintering aids were identified, i.e., Al_2O_3, La_2O_3 and MgO. Results for the first two will be discussed in detail.

The two phase diagrams available for the Y_2O_3 - Al_2O_3 system show conflicting information on the solubility of Al_2O_3 in Y_2O_3 and on the eutectic temperature between $Y_4Al_2O_9$ and Y_2O_3 (1940° to 1970°C). The present studies found Al_2O_3 solubility of ~ 0.02 mole % at a slightly lower eutectic temperature of 1930°C. Sintering below the eutectic resulted in 98% - 99% dense samples with increasing density and decreasing grain size as Al_2O_3 concentration was increased from 0.2 to 2.2 mole %. $Y_4Al_2O_9$ solid phase formed pinned grain boundaries, thus permitting more densification prior to the onset of secondary grain growth. Transparent samples were obtained by first employing an isothermal hold below the eutectic, and then increasing temperature to 2100°C for a 2 - 6 hr hold. This portion of the cycle accomplishes the final densification by the classical liquid phase sintering mechanism of solution-reprecipitation. Y_2O_3 dissolves in the thin grain boundary liquid phase at regions of high chemical potential and precipitates at regions of low chemical potential. Grain growth is again inversely proportional to the Al_2O_3 concentration, but in this case it is due to the longer diffusion distance through a thicker grain boundary film as Al_2O_3 concentration increases. Transmittance as a function of Al_2O_3 concentration through 1 mm thick discs is shown in Figure 1 (reflection losses are 18.7%). Low concentrations have too little Al_2O_3 to effect full densification. High concentrations exhibit excessive scattering from the second phase. Note that specular transmittance falls continuously from the optimum level due to scattering. The optimum concentration strikes a balance between the densification processes and light scattering from the grain boundary phase. The final grain size is ~ 250 μm which is considered large for ceramics but, suprisingly, strength is not seriously degraded. This is probably due to the cubic structure and high degree of microstructural perfection.

La_2O_3 has approximately 15 mole % solubility in cubic Y_2O_3. Since pure Y_2O_3 undergoes a phase transition to hexagonal at 2300°C, a very useful two phase field developes for Y_2O_3 - La_2O_3 up to the solid solution limit between 1860° - 2300°C. The transition temperature into the two phase field depends on composition, and the concentration of second phase developed depends both on starting composition and temperature. Fortunately, the high temperature phase has a very small volume change; thus ceramic parts cycled through this transition do not crack. The second phase markedly lowers grain boundary mobility. This enables the pore removal sintering process to proceed to completion, as the pores remain on grain boundaries until they are annihilated by the usual solid state diffusion process. The temperature is then lowered into the single phase region of the diagram, and grain growth becomes rapid as shown in Figure 2. Grain growth was difficult to control for compositions near the phase boundaries, and mobilities approaching intrinsic rates were observed. Normal grain growth was observed for the preferred compositions sintered at 2150°C, and grain size could be controlled between 30 - 150 μm. Specular transmittance is a measure of the perfection of the sintering/annealing process, and is shown for two sintering temperatures in Figure 3. The high specular transmittance is an indication of nearly complete pore removal, lack of second phase and the isotropic structure. Total transmittance was, of course, higher, ranging from 35 to 80 percent.

Strength, elastic modulus, thermal conductivity, and thermal shock resistance have been determined for both compositions.

Arc tubes were readily fabricated using standard technology, and exhibited transmittance values in the same range as current Al_2O_3 technology.

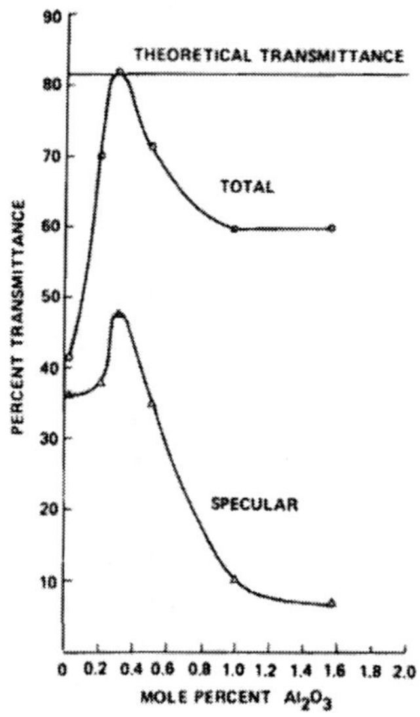

Figure 1. Transmittance of 1 mm Thick Y_2O_3 : (Al_2O_3)

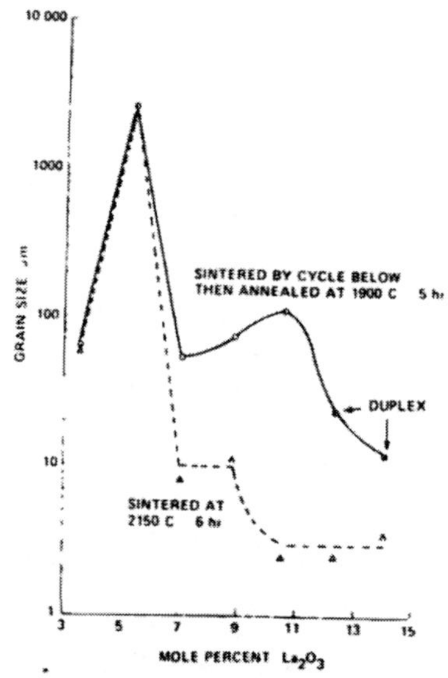

Figure 2. Grain Size After Sintering and Sintering/ Annealing Y_2O_3 : (La_2O_3)

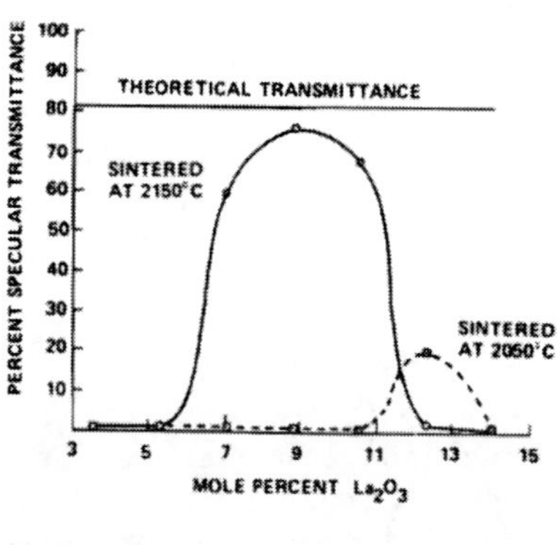

Figure 3. Transmittance of 1 mm Thick Y_2O_3 : (La_2O_3) Specimens

ANALYSIS OF HIGH PURITY METAL IODIDES FOR DISCHARGE LAMPS

L. Bruninx-Poesen and G. Solleveld
Philips Lighting Division
Eindhoven
The Netherlands

The quality of discharge lamps with metal iodide additives is strongly influenced by the degree of contamination of the compounds by oxygen and hydrogen. They can be present in different forms and it has proved necessary to develop a reliable method to analyse these elements quantitatively. This is performed by heating the specimen at high temperatures in a carbon crucible whereby all impurities containing hydrogen and oxygen are converted into hydrogen and carbon monoxide. These reaction products are transported using helium as a carrier gas into a gaschromatograph where they are separated on a molecular sieve column and determined quantitatively with a zirconia solid electrolyte detector. With this method concentrations of hydrogen from 1 to 1000 ppm (by weight) and oxygen from 10 to 10 000 ppm can be detected in samples of 10 mg.
Results of the analysis of some metal iodides from different origins are presented and discussed with regard to the influence of the impurities on the behaviour of discharge lamps.

NEW GETTERING ALLOYS FOR DISCHARGE LAMPS

by

E. Rabusin and M. Borghi

SAES GETTERS S.p.A., Via Gallarate 215, 20151 Milan, Italy

All lamps require a "getter" for cleaning up noxious gas impurities which are evolved within their jackets during life.

Modern high intensity discharge lamps (mercury, metal halide, sodium) and fluorescent tubes are especially sensitive to traces of hydrogen, water vapour and oxygenated gases which, when not quickly removed, rapidly deteriorate their performance and may cause premature failure.

In lamps with an evacuated jacket (especially low and high pressure sodium) the traditional barium flash getter is used whereas non-evaporable (bulk) getters are inserted within those having gas-filled jackets. Among these the Zr-Al alloy (St 101) is widely used (especially in metal halide lamps).

The very thin film of nitrides and oxides which covers the St 101 alloy grains and which prevents any reaction of the bulk of the getter with atmosphere gases, allows this material to be handled in open air without particular protection and to be subjected to the standard baking and exhaust processes of the lamp production without damage.

Hydrogen, water vapour, carbon monoxide and dioxide start to be sorbed, at a low speed, when the getter operation temperature exceeds about 300°C.

The sorption of these gases is very efficient when the getter temperature is between 350° and 450°C [(1)], which is a temperature that can be obtained when the getter is properly located.

Nitrogen is sorbed only if the temperature of the getter is in excess of 500°-550°C.

However, there may be some applications in which a lower operating temperature could be more desirable. Therefore some new zirconium based alloys have been developed which are able to sorb efficiently noxious impurities at temperatures starting from 200°-250°C. The sorption characteristics of these alloys are presented and compared with those of the St 101 alloy.

(1) Japn J. Appl. Phys. Suppl. 2, Pt. 1, 1974, 49-52

GASEOUS FLUORESCENCE AND VACUUM ULTRAVIOLET (VUV) EXCITATION

by
Jakob Maya
GTE LIGHTING PRODUCTS BUSINESS
100 Endicott St., Danvers, MA 01923

This paper attempts to show the viability of vapors and gases as alternatives to solid state phosphors in fluorescent lighting. First, the gas phase atomic and molecular processes producing visible radiation owing to UV excitation are described. Secondly, vapors and gases that can be exploited for fluorescence purposes in each category are discussed. A series of measurements on the absorption, emission processes and efficiency of fluorescence of some particular substances as gaseous phosphors are presented. Some of the areas of gaseous fluorescence that have been considered are:

1. Fluorescence as a result of photodissociation. Here a molecule (AB) dissociates as a result of absorbing a UV photon leaving one or two of the dissociated fragments in an excited state (A*) with subsequent emission of a visible photon. Schematically,

$$AB + h\gamma(UV) \longrightarrow A^* + B \quad \text{photodissociative excitation}$$
$$A^* \longrightarrow A + h\gamma(\text{visible}) \quad \text{emission}$$

The particular classes of molecules that have been studied are:

a - Diatomic metal halides
b - Triatomic metal halides

Table I shows quantum efficiency (QE) results of some of the metal halides studied. Brief experimental details on these measurements will be given.

TABLE I

Absolute quantum efficiency (QE) results, pressure, and temperature region of the measurements.

VAPOR	ABSOLUTE QE (%)	p* (torr)	T(K)	FLUORESCENCE REGION (nm)
TlI	42	0.03	624-728	377.6, 535.0
InI	20	0.013	521-693	410.2, 451.1
HgI_2	9	0.14	394-488	360-452
$HgBr_2$	26	0.22	393-467	320-508
CdI_2	9	0.06	513-646	360-660
ZnI_2	14	0.016	491-687	350-600

2. Fluorescence as a result of vacuum UV (VUV) excitation. Due to the high energy of VUV radiation, higher energy levels of atoms and molecules are accessible for excitation. This leads to fluorescence by new systems and/or increased efficiency of other fluorescing systems due to cascading and excitation of more than one radiative electronic state. Figure 1 shows fluorescence spectra of $HgCl_2$ excited by(VUV)part of deuterium radiation.

3. Fluorescence from complex metal halide vapors. The sensitive transition of the lanthanide earth metal ions in lanthanide earth halide compounds is exploited to obtain gaseous fluorescence. The enhanced volatility by complex formation is utilized to obtain fluorescence at relatively lower temperatures. Data on $TbCl_3$-$AlCl_3$ is presented. Figure 2 shows a typical configuration of a double chamber fluorescent light bulb utilizing a gaseous phosphor. The inner chamber is made of high UV transmission material. The combination of gaseous and solid phosphor is discussed. Applications of photodissociation and gaseous fluorescence in present HID lamps will be illustrated.

FIG. 1

VUV EXCITED SPECTRA OF $HgCl_2$. T = 150°C

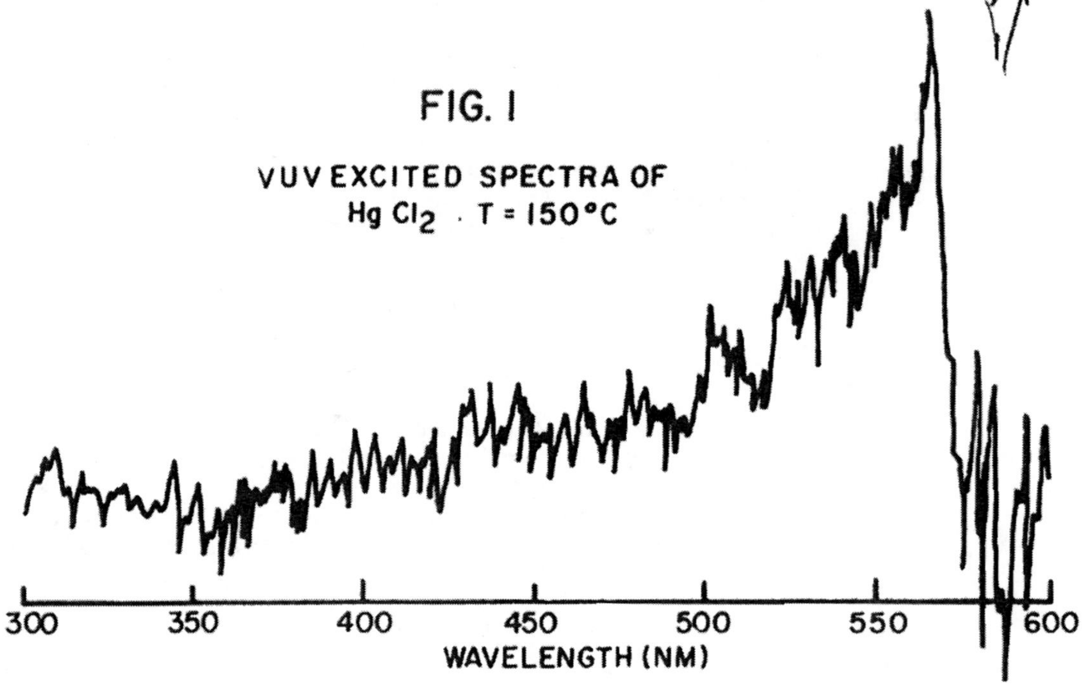

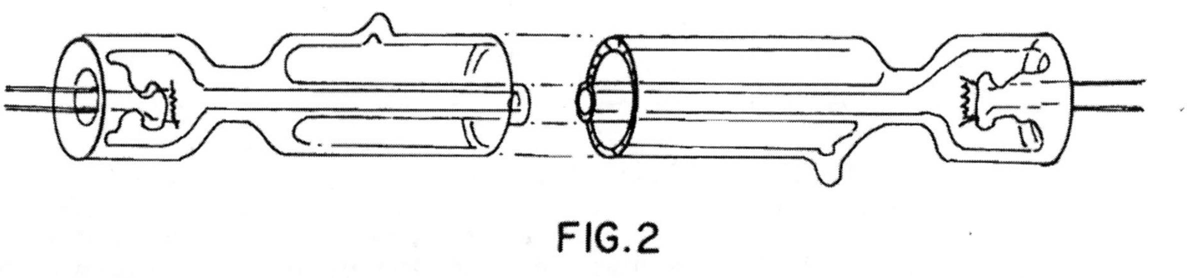

FIG. 2

PROGRESS IN SAPPHIRE ENVELOPE LAMP FABRICATION AND APPLICATIONS

44.

L. Reed, N. Anderson, R. McRae, and J. Richter
ILC Technology, Inc., Sunnyvale, California, USA.

Ceramics, most notably alumina (Al_2O_3) were first sintered to dense, pore free forms in the 1950-1960 period[1], and their potential for use as corrosion resistant lamp bodies was recognized soon after, leading to their use in the high pressure sodium arc lamp[2]. During this same period, the use of the single crystal alumina, i.e., sapphire, was explored both for UV flashlamp envelopes[3] and for military IR search light lamps[4] because of its superior in-line transmission in both the UV and the IR. Since that time, the use of sapphire as an envelope or window material has been developed by us for a variety of short arc[5], long arc[6] and resistance heated lamp types in power ranges from 5 watts to 8 kW. These arc lamps use both rare gas and alkali metal fills. It is the purpose of this paper to discuss the fabrication, performance and use of these lamp types.

Of critical importance in sapphire envelope lamp designs is the sapphire-to-metal seal. In general, techniques developed over the years for joining sintered alumina to metals in electron tube structures apply equally well to sapphire metal joining. These include seals made with glass solders or frits, seals made with molymanganese metallizing and subsequent brazing, and more recently developed active metal directly brazed seals[7]. However, lamp applications where operating temperatures of seals would be higher than in electron tubes and, in particular, where exposure to hot corrosive lamp fills such as alkali vapors would have to be tolerated, the conventional seals were found to be inadequate and new seals had to be developed.

For the sodium street lamp application, a corrosion resistant silica-free glass was developed that could be used to directly bond the niobium metal sealing ring and the translucent alumina envelope[2]. Because the lamp operates at high temperature in order to achieve significant vapor pressure of the metallic additives, a glass outer enclosure is used to prevent oxidation of the niobium. For other long arc alkali vapor lamp applications such as laser pumps[6] and infrared countermeasure radiators[8], it is advantageous to eliminate the outer envelope. To accomplish this an oxidation resistant active metal brazed seal using zirconium and the transition metals iron, nickel, and cobalt, was developed that has, in the case of the laser lamp, allowed life times of over 5,000 hours at seal temperatures approaching 700°C to be achieved in air. For these pumps the highest quality sapphire is used in order to minimize distortion of the arc image. In countermeasure lamps, lower cost "as grown" tubular sapphire suffic. A potassium fill is used in the laser pump lamp because the reversed resonance line radiation from the discharge very efficiently excites the Nd:YAG laser ma ial. A cesium fill is used for the pulsed countermeasure lamp for the most efficient production of psuedo-blackbody radiation in the mid-infrared.

Conventional short arc and long arc lamps with sapphire envelopes use nob gases as fill gases. With these fills seals made with molymanganese metallizi and copper brazing are adequate.

The tubular xenon short arc lamp[5] optical system offers advantages in ap plications where an exact relationship must be maintained between the arc hot reference mounting flange and external optics and/or reflector system. A mass tungsten-copper anode structure can be used for conductive cooling, thus elimi ing the need for forced air cooling. Applications requiring precision pre-foc lamps where small size and/or quiet operation are desirable include movie proj tion, interior spotlights, handheld searchlights, optical printers, as well as covert electro-optical monitoring, guidance and control systems. As a modific tion for special applications, the impervious nature of sapphire permits the addition of hydrogen to the xenon fill thus producing a higher voltage gradien more uniform brightness, and increased average brightness without a decrease o efficacy or a significant change of the spectral energy distribution.

Recently a water cooled sapphire envelope dc krypton lamp has been tested and found to be capable of handling powers at least as high as fused quartz en velope lamps. After testing it was discovered that very little deposition of material on the envelope adjacent to the electrodes had occurred, even after 5 hours of operation; it is speculated that the absence of dissolved water and o in the sapphire eliminates tungsten water cycle carrier reactions that apparen occur in fused quartz lamps.

Resistance heated lamps include those using graphite heaters and tungsten filaments where significant continuum radiation intensity is desired to the sa phire cutoff wavelength of approximately 6 microns.

Other materials such as beryllia, zinc sulfide, and magnesium fluoride ha been used as envelope materials for special applications where extreme chemica inertness and/or radiation further into the IR is desired.

REFERENCES

1. R. Coble, "Transparent Alumina and Methods of Preparation", U.S. Patent No. 3,026,210.

2. R.E. Hannem, W.C. Louden, C.I. McVey, "Thermodynamic and Experimental Studies of the High Pressure Sodium Lamp", IES Nat. Tech. Can Sept. 1968 preprint No. 18.

3. L.S. Nelson, D.A. Ramsay, "Flash Photolysis Experiments with a Sapphire Flashlamp", J. Chem. Phy. Vol 25, No. 2, p 372, 1956.*

4. J. Richter, "Development of a Compact Sealed Beam Plasma Arc Illuminator with an Integral Reflector", U.S. Army Contract DAAK02-71-C-0199.

5. TX Series "Sapphire Tubular Envelope Lamps", ILC Data Sheet 1970.

6. N.C. Anderson, R.M.F. Linford, J.M. Heitman, "Potassium Lamp for Space Laser Communications", SPIE Vol. 150, pp 48-54, Laser and Fiber Optics Communications, 1978.

7. L. Reed, "Ceramic to Metal Sealing", Electronics Ceramics Special Publication No. 3, Am. Cer. Soc. 1970.

8. "Pulsed Cesium Arc Discharge Lamp", L-3383, ILC Data Sheet.

*Also see G.P. Spindler, "Technique of Sealing Glass to Sapphire", 4th Symposium, The Art of Glassblowing, American Scientific Glassblowers Soc. May 1959.

DESIGN CONSIDERATIONS FOR DEVELOPING HPS LAMPS WITH IMPROVED COLOUR RENDERING

BY M.A. PARROTT, B. PRESTON AND R. HALL

It is known that by increasing the sodium vapour pressure, the colour rendering properties of HPS lamps can be greatly improved[1,2]. The optimum design of lamp will be that which achieves a tolerable CIE general colour rendering index (R_A), say 65 or above, and sacrifices least in lamp luminous efficacy and life.

In a typical standard lamp, optimised for maximum efficacy[3], the sodium vapour pressure is of the order of 100 torr, giving rise to a self reversal width for the sodium D lines of roughly 10 nanometers. Figure 1 shows that for such a lamp, as the sodium vapour pressure is increased, as measured by the D line separation, the R_A increases rapidly. R_A exceeding 60 for D line separations in excess of 30nm. The increase in sodium vapour pressure is achieved by increasing the temperature of the coolest part of the tube. The higher cool spot temperature will almost certainly result in some increase in wall loading (W/cm^2) in a practical lamp. The overall effect of producing high R_A lamps is to subject the PCA tubing and end seals to high sodium pressures at increased temperatures.

In our laboratory, we are currently investigating the relationship between colour rendering and efficacy, and increased sodium vapour pressure and wall

FIGURE 1

R_A AS A FUNCTION OF SELF-REVERSAL WIDTH FOR A LAMP CONTAINING AMALGAM OF .686 SODIUM MOLE FRACTION

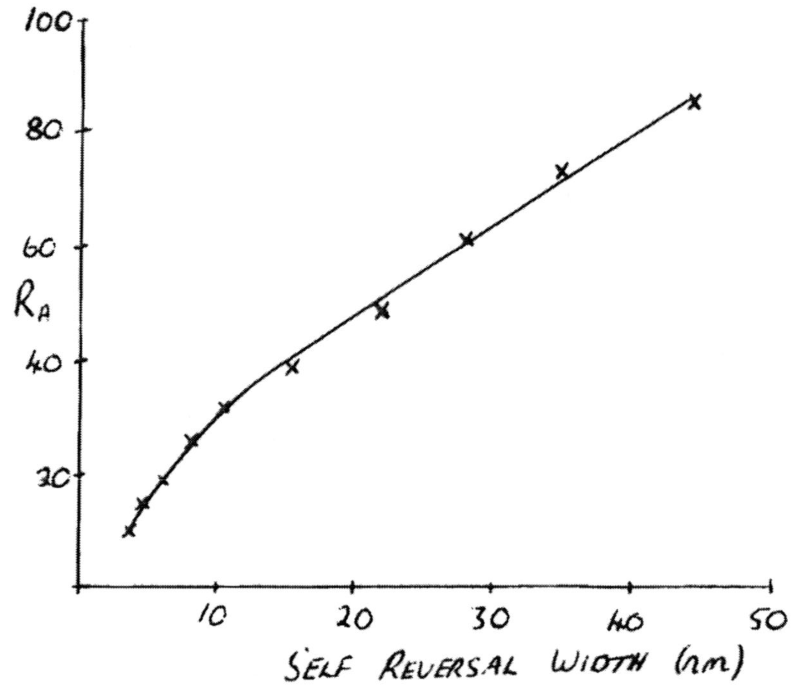

loading with life. The work is aimed at gathering design data for lamps of low power ratings i.e. 250W and less.

Experiments are designed to examine the relationships between R_A, source colour and luminous efficacy, when lamp parameters such as arc tube bore, arc current, cool spot temperature, sodium amalgam ratio and xenon fill pressure are varied. In these experiments, the cool spot temperature is varied by immersing one end of the lamp in a bath of molten indium of known temperature. The electrical characteristics, D line separation, spectral power distribution and the directional intensity of the lamp luminous output are measured. From the measured directional intensity, we deduce the lamp luminous output[4], and from the spectral power distribution, we are able to compute source colour and lamp colour rendering properties.

In a series of experiments, lamps with bores 4.0, 5.3 and 6.4mm and arc length of 29mm containing amalgam of .686 sodium mole fraction were measured at cool spot temperatures between 650 and 850 C, and currents between 0.5 and 2.0 amps. The measured variation of R_A with cool spot temperature for the arc tube of 5.3mm bore is shown in Figure 2. R_A was found to increase with increasing arc tube bore at constant cool spot temperature, as illustrated by the data contained in Table 1. Unfortunately, lamp efficacy falls more rapidly, at constant R_A, as the arc tube bore increases. This is shown clearly in Figure 3, a plot of relative lamp luminous efficacy as a function of R_A for the three arc tube bores.

Bore	R_A	Tc °C
6.4	65	780
5.3	65	795
4.0	65	810

TABLE 1

The highest correlated colour temperature observed was 2300°K. This compares with a value of 1950°K obtained for optimised tubes of similar dimensions.

Work on the durability of the PCA tubing and lamp seals is based on the examination of life tested lamps from two groups of experiments. In the first group of experiments, lamps of varying sodium vapour pressure and seal temperature, but having a fixed wall loading (W/cm^2), are constructed. In the second group of lamps, the sodium vapour pressure and end seal temperature are fixed and the wall loading varied. The required variations are acheived by careful choice of electrode backspace and electrode heat shield geometry.

FIGURE 2

R_A AS A FUNCTION OF COOL SPOT TEMPERATURE FOR A LAMP OF 5.3MM BORE, AMALGAM OF .686 SODIUM MOLE FRACTION, AT 1.0 AND 1.5A

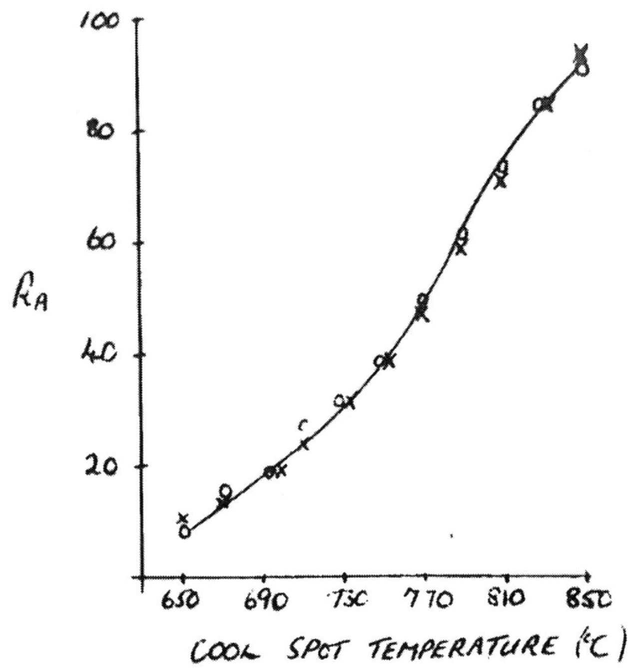

Initial observations indicate that the major loss of sodium is in the formation of beta-alumina in the form of dendritic growth. The position in the arc tube at which this occurs and the amount of growth is critically dependent upon wall loading and temperature distribution.

FIGURE 3

RELATIVE EFFICACY AS A FUNCTION OF R_A FOR LAMPS OF VARIOUS BORES, CONTAINING AMALGAM OF .686 SODIUM MOLE FRACTION

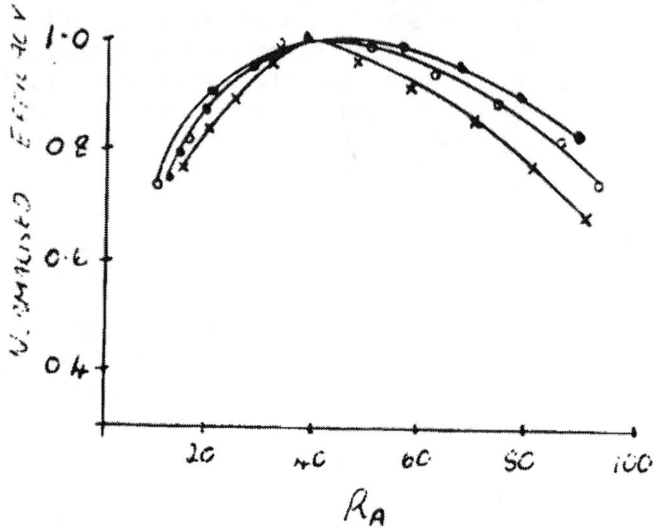

References

1) M. Mizuna, H. Akutsu and Y. Watarii, "New High Pressure Sodium Lamp with Higher Color Rendition". CIE XVII Session (Barcelona) 1971.

2) R.S. Bhalla, D.A. Larson and M.C. Unglert, "High Pressure Sodium Lamp with Improved Color Rendering". Paper presented at the Annual I.E.S. Conference 1978.

3) P.L. Denbigh, "Experimental Approach to High Pressure Sodium Lamp Design". Ltg. Res. and Technol. Vol. 10, No. 1 1978.

4) P.L. Denbigh. Ltg. Res. and Technol., 6, 62 (1974).

MAP N°1

ENSCHEDE – TOWN CENTRE

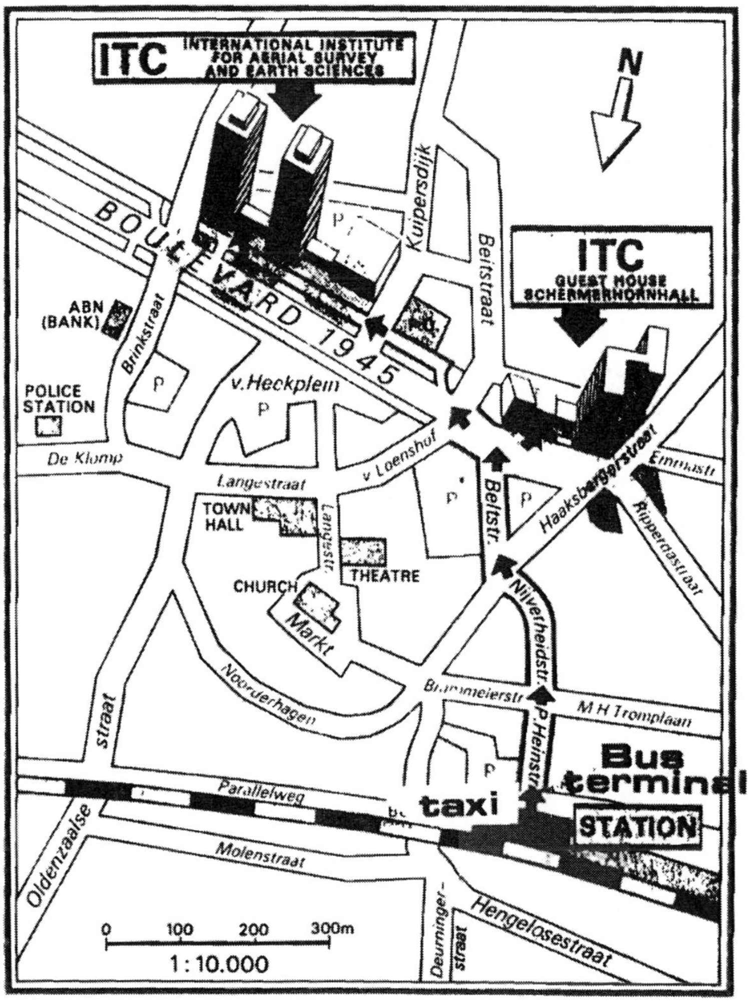

Orientation map for ITC visitors

PHYSICAL PHENOMENA ON ELECTRODES AND NEAR ELECTRODE REGIONS IN LOW PRESSURE GAS DISCHARGE LAMPS

E.V. Okhonskaya, Mordovian State University named after N.P. Ogarev, Saransk, U.S.S.R.

Under the inert gas pressure of some Torr thermal evaporation plays the main role in the oxide layer destruction when the electrodes operate in the stationary regime. Oxide loss rate reduction due to BaO and Ba diffusion return can be calculated by the following equation (1)

$$\frac{j_a^{gas}}{j_a^{vac}} = \frac{1}{1+\frac{3}{4}\frac{r_c}{\lambda}\ln\frac{r_t}{r_c}},$$

where j_a^{gas}, j_a^{vac} denote the density of evaporated molecules BaO and Ba in the inert gas and vacuum respectively; λ is the average free path of Ba in the inert gas. r_c, r_t are radii of the electrode and the discharge tube respectively. According to the calculations the oxide loss rate under 1 Torr argon pressure is reduced 10 times approximately.

The operation of electrodes in low pressure gas discharge lamps within the cathode half cycle is characterized by the presence of the positive potential gradient electric field in the electrode space. The rate of ionization of part of Ba atoms evaporated from th cathode in the immediate vicinity of the cathode spot within the cathode half cycle equals to $z_i = \alpha_i' n_a$, where α_i' - is the probability of Ba atoms ionization. The influence of the Ba returning cycles on oxide lossrate can be calculated from the equation (2)

$$\frac{j_r(r_o)}{j_a(r_o)} = \frac{X_2}{X_1} \quad (2); \quad X_1 = \int_{r_o}^{\infty} \frac{e^{-\psi(r)}}{r} dr; \quad X_2 = \int_{r_o}^{\infty} \frac{e^{-\psi(r)}}{r_o} \cdot \frac{K_1(\beta r)}{K_1(\beta r_o)} dr;$$

where K_1 is McDonald function of the primary order, $\psi(r) = \int_{0}^{r} \frac{eE(r)}{T_i} dr$; $\beta = \sqrt{\frac{\alpha_i}{D_a}}$; r_o, r - the external boundary of Langmuir layer and the current coordinate respectively, $E(r)$ - field strength in the cathode space; T_i is ion temperature, D_a - the coefficient of oxide atoms diffusion.

The calculations show that the accelerating field is able to secure the return of about 100% evaporated oxide atoms to the cathode within the cathode half-cycle. Consequently, the role played by the returning cycles of Ba ions in the reduction of oxide loss rate is more significant as campared to the diffusional return due to inert gas. The ions return of alkali-earth metals (AEM) prevent the reduction of donors' quantity in the oxide surface layer, preserve the supply of free Ba in oxide layer and stabilize emission characteristics.

In the anode half-cycle returning cycles of AEM ions are missing. Oxide loss is determined by its temperature dependent on the electrode construction, discharge current and ability of emission. Anode half-cycle plays the main role in oxide loss.

For the evaluation of electrode operation regime the following system of equation is used. It connects the conditions of AEM ions and atoms transfer with the field characteristics in the zone of ionization and the gas medium.

$$\bar{j}_r(0) = \bar{j}_a(0) + \bar{j}_i(0); \quad j_a = -\mathcal{D}_a \, \mathrm{grad}\, n_a; \quad \mathrm{div}\, \bar{j}_a = -z_i;$$

$$\bar{j}_i = -\mathcal{D}_i \, \mathrm{grad}\, n_i + b_i n_i \bar{E}; \quad \mathrm{div}\, \bar{j}_i = -\mathrm{div}\, j_a = z_i,$$

where $j_r(0)$ is the vector of AEM atoms flow density, finally removed from the oxide layer, $j_i(0)$ - the vector of ions current density brought back to the cathode by the electric field; $\mathcal{D}_i$, b_i, - diffusion coefficient and ions mobility in the inert gas, n_a, n_i - AEM atoms and ions concentration.

The given system permits:

1) to calculate $n_a(r)$, $n_i(r)$;

2) to determine the ratio between the AEM ions flow density directed to the cathode and atoms flow density evaporated from the cathode;

3) to calculate AEM atoms loss rate taking into account the returning cycles.

Having at our disposal the values $j_r(0)/j_a(0)$ and j_a^{vac}/j_a^{gas} for any possible electrode operation regime it is possible to determine the quantative values the average emission materials loss rate taking into account the role of cathode and anode half-cycles. Knowing the supply of BaO on electrodes it is possible to evaluate the lamp life at its permanent burning, taking into account the prices of switching on in the starter and starterless circuit, to determine lamp life in the real exploitation regime.

In conclusion the author express the gratitude to Rokhlin G.N. (Dr.S., Prof.) and Reshenov S.P. (M.Sc.) for their kindness help.

8. BACKGROUND INFORMATION; TECHNISCHE HOGESCHOOL TWENTE

Very shortly after the second world war it became obvious that the growing interest in scientific education made it neccesary to establish a number of new universities of technology. Since then in Twente, there has been a strive to establish such an institution there. On the 26th february 1953 the decision to establish a second university of technology fell in favour of Eindhoven as its place of location. However, the then present Minister of Education, Arts and Science - Mr. J.M.L.Th. Cals declared indeed that, from his points of view, the third university of technology should be established in time in Twente.

In the years 1957 up until 1960 inclusive, different towns sought nomination as the place of location for the third university of technology after the Piekaar and Neher commissions, amongst others set up by the Minister, had recommended that two new institutions should be established. The largest new institution, namely a university having a legal, an economic, a sociological and a technical faculty was thought of for establishment in the cast of the country. On the 16th of july 1959, the provincial government of Overijssel unanimously chose Twente. In deviation from the advice of the joint Piekaar/Neher commission who had selected Deventer as opposed to Drienerlo, Mr. Cals introduced a bill in parliament in which he proposed the country estate of Drienerlo as the location for the establishment of the university of technology in Twente. Two days later the country estate of Drienerlo was presented to the state by Enschede council. The law authorising the establishment of the third university of technology was ratified on the 1st of december 1961.

THE GROUND

The country estate of Drienerlo, occupying an area somewhat in excess of 98 hectares, served to realise the plan destined to be Drienerlo 1.

The famous architects ir.W. van Tijen and prof.ir. S.J. van Embden wererequested to plan the plot. But it was dr.ir. H.A. Stheeman, the dynamic governor, who put his stemp on the new university.

The building plot consisted of a large park and of farm-land, the typical scenic landscape of Twente. Builders were careful to integrate their design in this scenery, and they would rather adapt their plans than hew down a tree. We are grately indebted to them for sparing the park and the natural vegetation to this extent.
The university of technology would, according to Anglo-saxon example, be set up on a residential system (campus). It was decided to assign the western part (park) as a residential area, and to situate the buildings for education and

research in the eastern area. The campus area would be situated where both areas bordered on each other, an area where the refectory, meeting places and the sports centre would serve to be realised. Since june 1962 the area has been extended to cover 140 hectares. The farms of mr Bos and mr Lutke Holzik remain spared in the new conception. In view of the fact that there existed a difference in height of 19.7 feet between the western and the eastern parts of the plot and of the existence of three small rivulets in this area, three lakes were dug via which these could run their course. All the lakes are interconnected by culverts and spill-ways. The new roads, lakes, sportfields, lawns and, not to say the least of, the buildings determine the new character of Drienerlo.

THE LANDSCAPE

Seen geologically the eastern Netherlands falls under the so-called sand-surfaced areas having their own characteristic of widely varying contours. It is a gently undulating, hilly landscape. The first inhabitants of this part of the country settled down where high and low grounds met on the edges of the damp rivulet valleys. The low ground along the rivulet was overgrown with alder woods. These woods were partially cleared and transformed into grassland which owed its fertility to the annual overflow of silt containing water, the grassland became pasture and the cattle were brought back daily therefrom to the stalls where the manure was mixed with pear and stored, and later to be used for fertilising the high ground which was to be cultivated into arable land. In this manner, and through the annual build-up of a compost or turf-manure, there came into existance the old-time pastureland surface feature, the ash trees. The pasturelands were surrounded by oak windbreaks.

Parts of the woods in the vicinity of the farm have remained spared. After clearance of the high ground, self-developing moorland came into being. In this manner there came into existance the 'Twentse' farmstead or scenic landscape which was characterised by an extremely varied use of the soil which was adapted to the many differences in heigth and the natural condition of the terrain. With the advent of artificial fertiliser much was changed, heath and wastelands were reclaimed to partly become grassland and partly arable land. Above all large areas were afforested. This occured principally through the efforts of Dutch industrialists who extended the forested areas again, thad had been substantially reduced over the years, by the planting of conifers. Pretty well all the rivulets were canalised. A subsidiary feature of the improved water drawinage is that earlier drying of the high ground occurs. The described development of the 'Twentse' landscape is, in the main adapted to the university ground. Despite the fact that the area is not traversed by natural rivulets, occupation of this area took place a long time ago. The old arable lands were already cultivated 600 to 800 years ago. A study of the old topographical maps teaches us that an important part of the total land surface of the plot stood as waste ground and heathlands for a long time. Only just toward the end of the last century has there been a change herein. A major part of the woods were afforested at the beginning of the present century.

The former country estate of Drienerlo has, as well as the country estate of Horstlanden, a beautiful park-like character. The rivulets Bolhaarsbeek, Drienerbeek and the so-called 'Derde beek' (Third rivulet) are of recent date. Their courses are not even traced on old maps; they were probably dug during the second half of the last century. In that part to the south of the Langenkampweg, very little of the original landscape has been maintained. The old lakes, the old groves of trees and the old park surrounding them, and the old remains of the woods date from the time when Drienerlo was privately owned. On the place where the carillon now stands, there once stood a tea house. In the eastern part, several lanes and the farm Drienerhorst are to be remembered together with the arable lands associated therewith still forming a part of the original landscape.

Surface area of:	The terrain (total).	146 hectares.
	The water.	11 hectares.
	The buildings.	12 hectares.
	Hardstanding.	19 hectares.
	Greenery.	104 hectares.
Length:	Roads etc.	12 Kms (7.6 miles).
	Footpaths.	8.5 Kms (5.3 miles).
Heigth difference from east to west		6 meters (19.5 feet).

THE PURPOSE OF THE UNIVERSITY OF TECHNOLOGY

The purpose of the university of Twente, which was opened during the first public meeting of the senate by Her Royal Majesty Queen Juliana on the 14th september 1964, varies in different respects from scientific education.

At the inception of the first study year 1977-1978 the number of students was approximately 2500 in total. In the years 1964 to 1967 inclusive, the law approving the establishment of the university was enacted and which law authorised the university to control the influx of new students to a maximum. In consultation with the Minister, the maximum determined number of students were granted entry by the drawing of lots.

The total number of students reporting themselves for the first study year, 1964-1965, totalled 234.

As at april 1978 the total number of engineers having graduated from Twente University was 1000

The number of co-workers (scientific body, technical and administrative services) totals approximately 1200.

MANAGEMENT AND TOP–STRUCTURE ORGANISATION.

As the result of universities administration reform act of 1970 the management of the university rests with the University Council and the Executive Committee.

The University Council is responsible for determining the general policy. It lays down the rules and regulations as well as determining the budgetary estimates, the financial plan and the development plans.

The University Council outlines directives for the organisation and co-ordination of education and scientific activities and is charged with the provisions for the students.

The executive Committee is charged with the day-to-day administration of the University; it is the policy preparing and executing body. The University Council and Executive Committee is advised in matters of education and scientific activities by the Faculty Council on which chairmen of departments and sub-departments sit as members thereof. The Chairman of the last named Council is the Rector Magnificus who represents the university in its intercourse with other universities.

The University Council comprises 27 members chosen by the university community and 6 external members appointed by the Crown after consultation with the University Council, appointed members, the Rector Magnificus and two members of the scientific body chosen by the University Council.

THE UNIVERSITY OFFICE OF ADMINISTRATION.

The secretary of the University is charged with the task of heading the office of administration. He also exercises the function of secretary to the executive committee. The office of administration comprises the cabinet (KAB), press and public relations department (PV), the departments for general and financial matters (AFZ) and the personnel department (PZ).

CDO.

The Centre for Didactic and educational research supports the educational departments in their educational duties.
Amongst others in the CDO, there is an audiovisual centre which has the task of advising and informing in matters concerning the use of audiovisual media and for the provision of audiovisual programs.

CAMPUS.

The explanatory memorandum in the statutory publication 1964 No: 324 authorising establishment of the University of Technology states– 'The University of Technology-Twente' will strive, linked by the different lines of development of University education, to build up an academic community wherein not only research and education as such, but also studies and student life, are integrated as much as possible. Should there be insufficient attention paid to any one of these fields then one would undergo as much a disadvantageous influence as another. This basic concept forms the background in which the 'Campus University' is set out in relief'. Since then the Twente University has not been by-passed by the social changes. On march 1st 1974, the installation of a new Campus administration took place. In the sectors of sport, culture, refectory, health care, accomodation and campus personnel representitives thereof took their places in this administrative body and, moreover, two external members also have places therein. At the same time a Director of campus facilities was appointed.

The campus must be understood to comprise, in its entirety, study and community life. Primarily 'De Vrijhof' accomodates the central library, the campus management, the campus Director, the campus administration, the office for student accomodation, the office of cultural affairs, the office of student psychology, the office for student-facilities, the joint pastorates and studium. There are different halls available for divine services, performances and exhibitions. Various student societies and committees are also accomodated in 'De Vrijhof' as is the Drienerlo Student Council (SRD) and the editorial staff of the opinion news sheet – 'De Wekker'.

The farm (the former refectory) now accomodates a team which works on a project itself known as 'The farm' and which project has, for its central theme, research into the field of energy and environment.

'De Bastille' (Refectory and central meeting building of the campus) and the sports centre also fall within the campus organisation. 'De Bastille' also acco-

Foundation for the Advancement of the Science & Technology of Light Sources
FAST-LS
Belmayne House
99 Clarkehouse Road
Sheffield, United Kingdom
S10 2LN

ISBN 978-1-5108-4080-5

CURRAN ASSOCIATES INC.
proceedings.com

18th International Scientific Conference on Current Problems of the Corporate Sector (ICRTEG 2021)

SHS Web of Conferences Volume 115 (2021)

Bratislava, Slovakia
20 May 2021

edp sciences